U.S. ARMY WEAPONS SYSTEMS 2009

DEPARTMENT OF THE ARMY

Foreword by William D. Wunderle, Lieutenant Colonel, U.S. Army

Skyhorse Publishing

Skyhorse Publishing books may be purchased in bulk at special discounts for sales promotion, corporate gifts, fund-raising, or educational purposes. Special editions can also be created to specifications. For details, contact the Special Sales Department, Skyhorse Publishing,
555 Eighth Avenue, Suite 903, New York, NY 10018 or info@skyhorsepublishing.com.

www.skyhorsepublishing.com

10 9 8 7 6 5 4 3 2 1

ISBN-13: 978-1-60239-336-3

Printed in Canada

Foreword

The United States Army evolved from George Washington's amateur Continentals, barely able to confront the British Army throughout the thirteen colonies, into the most highly sophisticated professional military force in the world, capable of conducting worldwide operations. In a world governed by geopolitical and economic interests, enemies change.[1] Like the American Revolution, the current conflict—the Long War—has many facets. It includes State and Non-State actors and a wide range of adversaries, all using diffuse technologies in a complex distributed battlespace.

The world has changed greatly in the years since September 11, 2001. Major engagements in Afghanistan and Iraq have significantly degraded al-Qaeda, with fledgling democracies emerging in both countries. Popular support for the group in the Muslim world is at its lowest point since 2001. However, the resurgence of the Taliban in Afghanistan, the recalcitrance of al-Qaeda cells in the Federally Administered Tribal Areas (FATA) of Pakistan, the return of foreign fighters to al-Qaeda franchises worldwide, and the rise of Hizballah, Jaysh al-Mahdi, the Special Groups, and other Iranian proxies, all suggest that there is still much work to be done. Many of America's adversaries and enemies have no national standing, no government, no uniformed military, no allegiance to international law, no interest in humanity, and no restrictions on the weapons they are willing to use to achieve economic, political, and theocratic domination of the world.

Recognizing this, the Pentagon's new U.S. National Defense Strategy, released in July 2008, emphasizes the need for the U.S. military and the entire government to be prepared to fight global terrorism and related small-scale conflicts like the wars in Iraq and Afghanistan.[2] The National Defense Strategy recognizes that the United States is in the midst of fighting the "next war" and places the Long War against extremism as the top priority of the U.S. military for the foreseeable future, above potential conventional challenges from China and Russia.

The Army's mission is to "fight and win our Nation's wars by providing prompt, sustained land dominance across the full range of military operations and spectrum of conflict in support of combatant commanders."[3] Army forces provide the capability—by threat, force, or occupation—to promptly gain, sustain, and exploit comprehensive control over land, resources, and people. One of the ways the Army accomplishes its mission is through the organizing, equipping, and training of its forces for the conduct of prompt and sustained combat operations on land. The role of the U.S. Army constantly changes to meet new threats and challenges. Today, the U.S. Army is in the midst of executing the most profound transformation since World War II. While, the "future" fight against global terrorism is ongoing, and will continue to be a top concern for the U.S. military and policymakers, the United States cannot afford to lose its conventional qualitative edge or accept technological inferiority.

U.S. Army Weapons Systems 2009 is a compendium of the most innovative and important U.S. Army weapons systems and initiatives. This fully illustrated manual focuses not only on specific U.S. Army weapons systems, but also on the U.S. Army's Future Combat Systems (FCS), simulations, and the Science and Technology Strategies (S&T) that will enable the Future Force while seeking opportunities to enhance the Current Force of the United States Army.

In the wake of September 11, 2001, the American public yearns to understand not only the nation's strategic purpose, but also its military. As you page through this book, you will see the combat systems, weapons, and equipment used by America's Army. What you will not see however, is the most important part of the equation—the men and woman of the U.S. Army. Despite being the most technologically advanced Army on the globe, it is the soldier who makes America's Army the most respected force in the world. As "it is the musician who creates the music, not the instrument; the surgeon who operates, not the scalpel,"[4] it is the American soldier, not the weapons, who makes the U.S. Army the most lethal army in history.

William D. Wunderle
Lieutenant Colonel
United States Army

1 For more insight into how the Army has changed over time, see Chester H. Hearn's, *Army: An Illustrated History*, Zenith Press, Saint Paul, Minnesota, 2006.
2 Secretary of Defense Robert M. Gates, Remarks to the Heritage Foundation (Colorado Springs, CO), May 13, 2008, http://www.defenselink.mil/speeches/speech.aspx?speechid=1240.
3 Headquarters, Department of the Army (HQDA). *FM 1, The Army*. Washington, DC: Department of the Army, June 2005, p 2-7.
4 Fred J. Pushies, *Weapons of Delta Force*, Zenith Press, Saint Paul, Minnesota, 2002.

Dear Reader:

The weapon systems and equipment described in this reference book represent an essential aspect of our commitment to the security of the nation, the preparedness of the Soldier, and the readiness of the Army.

We serve the Soldier, the centerpiece of our combat systems. Our Soldiers are critical to an Army that is serving the nation at war; more than 268,000 troops are answering the Call to Duty in more than 120 countries worldwide, and they stand ready to fulfill all current missions, including homeland security.

The Army is investing in recapitalizing and modernizing the Current Force to ensure continuing Army dominance in the face of emerging threats. New capabilities have been fielded to support current operations. New efforts are getting equipment to our Soldiers faster than ever. We are maintaining readiness and improving the capabilities of units returning from and preparing for deployment.

In addition, the Army continues to develop the Future Combat Systems (FCS) initiative, which represents the Army's first full-spectrum modernization in nearly 40 years. When fully operational, FCS will provide the Army and the joint force with unprecedented capability to see the enemy, engage him on our terms, and defeat him on the 21st century battlefield. FCS will become the face of the Future Force.

The following pages describe our investments in the successful acquisition and sustainment of weapon systems and equipment. As you use this informative resource, however, remember that even the most technologically advanced platforms are useless without the skill and dedication of the American Soldier. Working with Congress, we will strive to provide our Soldiers with the best possible equipment so that our Army will be ready to meet today's requirements and tomorrow's challenges.

Claude M. Bolton, Jr.

Claude M. Bolton, Jr.
Assistant Secretary of the Army
(Acquisition, Logistics and Technology)

Table of Contents

Table of Contents

How to Use This Book

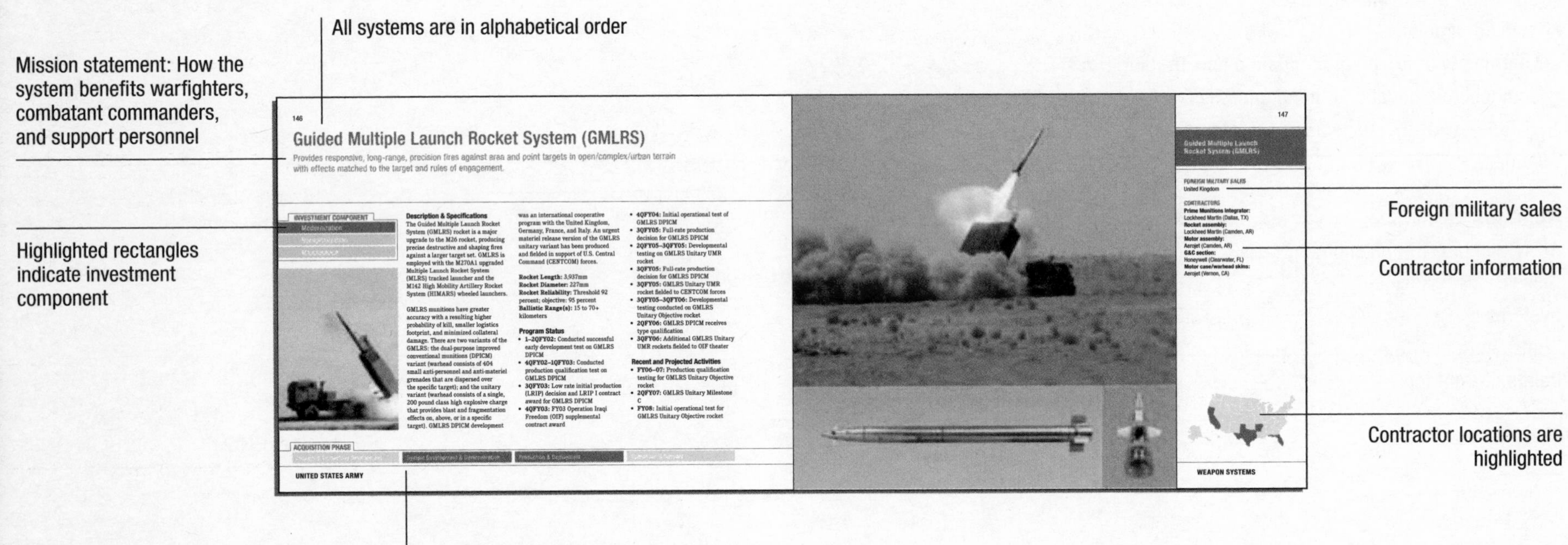

Highlighted rectangles indicate acquisition phase

About This Edition

THE CHANGES

Readers familiar with this publication will notice some changes this year that make United States Army Weapon Systems more intuitive, readable, and attractive. In this year's edition:

- An overview of the Army's Future Combat Systems (FCS) is included
- The systems are still in alphabetical order, and each has its own spread, but the information and the photos have been rearranged for greater clarity and ease of reference
- The Acquisition Phase and Investment Component are now indicated by a system of shaded rectangles on the left page for easy recognition
- Science & Technology is still presented in a separate chapter
- The panel on the right page contains critical system information, such as system name, foreign military sales, and contractor information, as well as a map showing contractor locations

For explanations of each of the elements on a typical system spread, see the example on the left.

WHAT ARE INVESTMENT COMPONENTS?

Modernization programs develop and/or procure new systems with improved warfighting capabilities.

Recapitalization programs rebuild or provide selected upgrades to currently fielded systems to ensure operational readiness and a zero-time, zero-mile system.

Maintenance programs include the repair or replacement of end items, parts, assemblies, and subassemblies that wear out or break.

For additional information and definitions of these categories, please see the Glossary.

WHAT ARE ACQUISITION PHASES?

Concept & Technology Development refers to the development of a materiel solution to an identified, validated need. During this phase, the Mission Needs Statement (MNS) is approved, technology issues are considered, and possible alternatives are identified. This phase includes:

- Concept exploration
- Decision review
- Component advanced development

System Development & Demonstration (SDD) is the phase in which a system is developed, program risk is reduced, operational supportability and design feasibility are ensured, and feasibility and affordability are demonstrated. This is also the phase in which system integration, interoperability, and utility are demonstrated. It includes:

- System integration
- System demonstration
- Interim progress review

Production & Deployment achieves an operational capability that satisfies mission needs. Components of this phase are:

- Low rate initial production (LRIP)
- Full rate production decision review
- Full rate production and deployment

Operations & Support ensures that operational support performance requirements and sustainment of systems are met in the most cost-effective manner. Support varies but generally includes:

- Supply
- Maintenance
- Transportation
- Sustaining engineering
- Data management
- Configuration management
- Manpower
- Personnel
- Training
- Habitability
- Survivability
- Safety, Information technology supportability
- Environmental management functions

Because the Army is spiraling technology to the troops as soon as it is feasible, some programs and systems may be in all four phases at the same time. Mature programs are often only in one phase, such as operations and support, while newer systems are only in concept and technology development.

FUTURE COMBAT SYSTEMS

FCS

The Army's Future Combat Systems (FCS), the Army's first full-spectrum modernization in nearly 40 years, is...

The face of the future force

When fully operational, FCS will provide the Army and the joint force with unprecedented capability to see the enemy, engage him on our terms, and defeat him on the 21st century battlefield.

FCS is a joint, networked system of systems, consisting of a network and 18 individual systems and using an advanced network architecture that will enable levels of joint connectivity, situational awareness and understanding, and synchronized operations previously unachievable. It is designed to interact with and enhance the Army's most valuable weapon—the Soldier.

FCS (BCT) is the Army's principal modernization strategy that is the embodiment of the modular force, a modular system designed for "full-spectrum" operations. It will network existing systems, systems already under development, and systems to be developed to meet the requirements of the Army's Future Force. It is adaptable to traditional warfare as well as complex, irregular warfare in urban terrains, mixed terrains such as deserts and plains, and restrictive terrains such as mountains and jungles. It can also be adaptable to civil support, such as disaster relief.

In 2014 the Army will begin fielding Brigade Combat Teams (BCTs), units that will operate using all of FCS's systems. When compared to current "heavy" brigades, the BCT offers numerous advantages.

The structure of modern warfare is changing

The very structure of modern warfare is changing. Today's warfare is increasingly irregular, requiring us to find, engage, and defeat the enemy on complex terrain. Adversaries are adapting and honing their skills to defeat our current strengths and abilities, using guerrilla and terror tactics to attack, disrupt and harass our forces. There is a need to transition away from the 20th century, Cold War model that relied on massive logistics buildup, heavy brigades, sequential operations, linear warfare, and intelligence gained by direct observation/contact. FCS mitigates the dilemma of irregular warfare by providing light, agile Brigade Combat Teams with a small logistics footprint that is networked and capable of conducting simultaneous operations; non-linear warfare that directly attacks enemy centers of control; and exploits intelligence gained via remote reconnaissance and surveillance.

At the heart of the FCS (BCT) is the network, which will allow every FCS system—from unmanned vehicles to precision weapons—to share information and work together. The network will facilitate decision-making not just at the brigade level, but all the way down to the battalion and company levels. The network is FCS's enabler, allowing the Army to achieve greater situational awareness, improved survivability, improved lethality, improved efficiency, and joint operability.

Meeting the Challenge

On today's battlefield, the availability of real-time information is critical for success. Up-to-the-second information often is the difference between leading an attack and reacting to the enemy's attack. FCS technology will allow our Soldiers to see first and understand first, from a position out of harm's way. Systems such as the Unattended Ground Sensor (UGS), Unmanned Ground Vehicles (UGVs) and Unmanned Aerial Vehicles (UAVs) will provide information about the enemy's position in individual buildings and neighborhoods, as well as over the horizon. This information will be fed into the network and immediately shared with brigade, battalion, and company commanders, and even platoon leaders. This networked surveillance increases certainty of information and reduces tactical risk to our Soldiers. In summary, FCS provides enhanced situational awareness.

On today's battlefield, precision weapons are necessary to defeat enemies who are often intermixed with civilian populations or deeply imbedded in restrictive terrain such as mountainous regions. FCS systems such as the Mounted Combat System (MCS), Non-Line of Sight-Cannon (NLOS-C) and Non-Line of Sight-Mortar (NLOS-M), combined with FCS's unmanned systems and our Soldiers, provide the ability to destroy enemy concentrations as well as enhanced ability to identify combatants interspersed with non-combatants and to engage with precision munitions that reduce the risk of collateral damage or unintended consequences. In summary, FCS increases lethality.

FCS increases survivability, lethality, and efficiency

On today's battlefield, Soldiers in complex environments are at risk within vehicles, due to the enemy's use of improvised explosive devices (IEDs), rocket-propelled grenades (RPGs), and anti-tank missiles, and on foot, as they navigate complex terrain where the enemy is well hidden and traditional fighting vehicles are largely ineffective.

FCS reduces that risk by using unmanned vehicles such as the Armed Robotic Vehicle (ARV), Small Unmanned Ground Vehicle (SUGV), and Multifunctional Utility/Logistics and Equipment Vehicle (MULE) to locate and engage the enemy, identify toxic chemicals, destroy tanks, and disable land mines. FCS manned vehicles are agile, and carry a full suite of hit avoidance technologies such as active protection systems (APS) which allow Soldiers to stay mounted longer, which reduces their risks, before delivering them close to the fight. In summary, FCS increases survivability.

On today's battlefield, it is imperative that we maximize the fighting capacity of our force. The FCS (BCT) features smaller, lighter vehicles which quickly transport more combat power to where it's needed. In addition, FCS vehicles will require much less fuel, reducing the number of refueling vehicles. By building many of its systems on a common chassis, the number of mechanics and spares will be reduced. In addition, reduced support requirements mean fewer convoys. Threats from IEDs will be minimized by FCS's sensors and robots. In summary, FCS increases efficiency and reduces the Army's logistics footprint, resulting in fewer support Soldiers and vehicles, thus saving lives and money.

On today's battlefield, it is important that joint forces and allies are able to work together to defeat a common enemy. The FCS (BCT) is designed to act as a unified combined armed force in the joint environment. FCS will benefit all ground forces, including the Marines, multipurpose, and special operations, through spin outs of FCS technology. In summary, FCS supports the joint environment.

FCS will allow the Army to find, fix, and finish the enemy on the 21st century irregular battlefield

FCS is a System of Systems

Because FCS is a system of systems, the whole is more than the sum of its parts. By linking the capabilities of 18 cutting-edge systems with a state-of-the-art network and the unmatched abilities of the American Soldier (18+1+1), FCS will be the fulfillment of the modular force, providing a joint, full-spectrum approach to warfare that will allow the Army to find, fight, and finish the enemy on the 21st century, irregular battlefield.

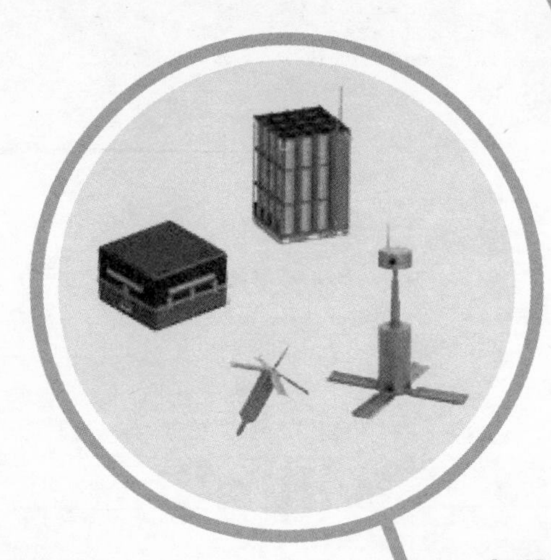

(18+1+1) Platform & Sensors

Manned Ground Vehicles

Mounted Combat System (MCS)

The Mounted Combat System (MCS) provides Line-of-Sight (LOS) and Beyond-Line-of-Sight (BLOS) offensive firepower capability allowing BCTs to close with and destroy enemy forces. The MCS delivers precision fires at a rapid rate to destroy multiple targets at standoff ranges quickly and complements the fires of other systems in the BCT. It is highly mobile and maneuvers out of contact to positions of advantage. It is capable of providing direct support to the dismounted infantry in an assault, defeating bunkers, and breaching walls during the tactical assault. The MCS also provides BLOS fires to destroy point targets through the integrated sensor network. This capability enhances SoS lethality and significantly increases the options available to the BCT commander for the destruction of point targets through the integrated fires network. MCS shares a common chassis with the other FCS Manned Ground Vehicles and consists of LW 120mm Cannon and an Ammunition Handling System. XM307 is a secondary armament to be used on the MCS.

Infantry Carrier Vehicle (ICV)

The Infantry Carrier Vehicle (ICV) consists of four platform versions: a Company Commander; a Platoon Leader; a Rifle Squad; and a Weapons Squad. All four platform versions appear to be identical from the exterior to prevent targeting of a specific Infantry Carrier Vehicle (ICV) variant type (e.g., Platoon Leader). The Infantry Platoon includes an ICV Platoon Leader variant; three ICV Rifle Squad variants; and an ICV Weapons Squad variant. The ICV Rifle Squad variant and ICV Weapons Squad variant each deliver 9-person infantry squads to a location from which they will conduct a close assault. The ICV will effectively employ weapon systems and rapidly maneuver during blackout, day and night operations, inclement weather, and limited visibility periods. The ICV carries the majority of equipment freeing the individual Soldier to focus on mission. The squad will have access to Army and Joint fire delivery systems from external sources to provide extended range, networked responsive precision or volume fires on demand in support of tactical maneuvers. The ICV can move, shoot, communicate, detect threats, and protect crew and critical components under most land-surface environments. Data transfer with other components of the BCT permits constant update of the common operational picture and rapid identification of targets.

Non-Line of Sight-Cannon (NLOS-C)

The Non-Line of Sight-Cannon (NLOS-C) is an indirect fire support component of the System of Systems (SoS) of the FCS (BCT). It will be organic to and provide networked, extended-range, responsive and sustained precision attack of point and area targets in support of the FCS (BCT). It fires a suite of munitions that include special purpose capabilities to provide

Real Systems Real Success

Autonomous Navigation System (ANS)/ Stryker Leader/Follower Demonstration Fall 2005

Software build 1 testing in the System of Systems Integration Laboratory (SoSIL) Summer 2006

a variety of effects on demand including precision guided munitions such as the XM982 Excalibur. NLOS-C will provide close support and destructive fires for tactical standoff engagement during both offensive and defensive operations in concert with line-of-sight, beyond-line-of-sight, other NLOS, external and joint capabilities in combat scenarios spanning the spectrum of ground combat and threats.

The NLOS-C will be a self propelled howitzer with a two man crew. It will have a 155 mm, Zone 4, 38 caliber cannon, fully automated armament system and a high level of commonality with other MGV variants. It will mount the XM-307 25mm Advanced Crew Served Weapon (ACSW) as its secondary armament and will incorporate a suite of protection measures to enhance crew and platform survivability. The NLOS-C will be deployable worldwide and will operate in a wide range of natural environmental conditions. The cannon will be able to move rapidly, stop quickly, and deliver lethal first round effects on target in record time. The NLOS Cannon will have a multiple round simultaneous impact (MRSI) capability. The MRSI capability, coupled with the NLOS-C superior sustained rate of fire, will provide record effects on target from a smaller number of systems. The cannon, like all Manned Ground Vehicle (MGV) variants, can rapidly rearm and refuel, and its system weight makes it uniquely deployable. Fully automated handling, loading, and firing will be another centerpiece of the NLOS-C. The NLOS-C balances deployability and sustainability with responsiveness, lethality, survivability, agility, and versatility. The NLOS-C will be designed to minimize its logistic and maintenance footprint in the theater of operation and to employ advanced maintenance approaches to increase availability and to support sustainability.

Non-Line of Sight-Mortar (NLOS-M)

The Non-Line of Sight-Mortar (NLOS-M) is the short-to-mid-range indirect fire support component of the System of Systems (SoS) of the FCS (BCT). It will be organic to and provide networked, responsive and sustained indirect fire support to the Combined Arms Battalion (CAB) in the FCS Brigade Combat Team (BCT). It fires a suite of 120mm munitions that include special purpose capabilities to provide a variety of fires on demand including precision guided munitions such as Precision Guided Mortar Munitions (PGMM). NLOS-M will provide close support and destructive fires for tactical standoff engagement during both offensive and defensive operations in concert with line-of-sight, beyond-line-of-sight, other NLOS, external and joint capabilities in combat scenarios spanning the spectrum of ground combat and threats.

The NLOS-M will mount the MGV Advanced Crew Served Weapon (ACSW) as its secondary armament and will incorporate a suite of protection measures to enhance crew and platform survivability. The Section Chief and Driver will occupy the vehicle's Common Crew Station while the remaining crew member will sit immediately to the rear. The current design has the crewmen facing forward during movement and turning their seats rearward when the NLOS-M is emplaced and ready to fire. The primary duties of the rear crewman is to stow ammunition during rearms, prepare the 120mm mortar ammunition for firing by removing and stowing unneeded propellant charges setting the fuze, and inserting the prepared round into the loading device. The semi-automated ammunition handling on the NLOS-M will present the proper round from the magazines to the crewman, after the crew has prepared the ammunition for firing and put it in the loading elevator, load and fire the round. The automated fire control will compute firing data and point the tube.

The NLOS-M will be deployable worldwide and will operate in a wide range of climatic conditions. The NLOS-M will have a high level of commonality with other MGV variants and will be designed to minimize its logistic and maintenance footprint in the theater of operation. The NLOS-M will employ advanced maintenance approaches to increase availability and support sustainability.

Reconnaissance and Surveillance Vehicle (RSV)

The Reconnaissance and Surveillance Vehicle (RSV) features a suite of advanced sensors to detect, locate, track, classify and automatically identify targets from increased standoff ranges under all climatic conditions, day or night. Included in this suite are a mast-mounted, long-range electro-optic infrared sensor, an emitter mapping sensor for radio frequency (RF) intercept and direction finding, remote chemical detection, and a multifunction RF sensor. The RSV also features the onboard capability to conduct automatic target detection, aided target recognition and level one sensor fusion. To further enhance the scout's capabilities, The RSV is equipped with unattended ground sensors (UGS), a Small Unmanned Ground Vehicle (SUGV) with various payloads and two unmanned aerial vehicles (UAVs).

Command and Control Vehicle (C2V)

The Command and Control Vehicle (C2V) is part of the family of manned ground vehicles and is the hub of battlefield command and control. The C2V platform provides the tools for commanders to synchronize their knowledge of combat power with the human dimension of leadership. It is located within the headquarters sections at each echelon of the BCT down to the company level, and with the integrated command, control, communications, computers, intelligence, surveillance and reconnaissance (C4ISR) suite of equipment, the C2V provides commanders with the ability to command and control on the move.

Via mission workstations, C2Vs contain the interfaces that allow commanders and their staffs to access Battle Command mission applications including: mission planning and preparation, situation understanding, Battle Command and mission execution, and warfighter-machine interface. These applications enable commanders and their staffs to perform tasks such as fusing friendly, enemy, civilian, weather and terrain situations and distributing this information via a common operating picture. Commanders also utilize the C2V's integrated C4ISR suite to receive, analyze and transmit tactical information both inside and outside the BCT. The C2V can also employ unmanned systems, such as unmanned aerial vehicles to enhance situational awareness throughout the BCT and is slated to use XM 307 ACSW as secondary armament.

Medical Vehicle-Treatment (MV-T) and Evacuation (MV-E)

The Medical Vehicle is designed to provide advanced trauma life support within one hour to critically injured Soldiers. The Medical Vehicle serves as the primary medical system within the Brigade Combat Team (BCT) and will have two mission modules: Evacuation and Treatment. The time-sensitive nature of treating critically injured soldiers requires an immediately responsive force health protection system with an expedient field evacuation system. The Medical Vehicle-Evacuation (MV-E) vehicle allows trauma specialists, maneuvering with combat forces, to be closer to the casualty's point-of-injury and is used for casualty evacuation. The Medical Vehicle -Treatment (MV-T) vehicle enhances the ability to provide Advanced Trauma Management (ATM)/ Advanced Trauma Life Support (ATLS) treatments and procedures forward for more rapid casualty interventions and clearance of the battlespace. Both Medical Vehicle mission modules will be capable of conducting medical procedures and treatments using installed networked telemedicine interfaces, Medical Communications for Combat Casualty Care (MC4) and the Theater Medical Information Program (TMIP).

FCS Recovery and Maintenance Vehicle (FRMV)

The FRMV is the recovery and maintenance system for employment within both the Brigade Combat Team (BCT) and divisions and contributes to sustaining and generating combat power to the Future Force

Non-Line of Sight-Cannon (NLOS-C) Demonstrator firing at Yuma Proving Ground
Fall 2006

Manned Ground Vehicle Common chassis Hybrid Electric Drive and Band Track Demonstrator at Aberdeen Proving Ground (APG)
Fall 2005

FCS surrogate network vehicle at Joint Expeditionary Force Experiment (JEFX06)
Summer 2005

structure. Each BCT will have a small number of 2-3 man Combat Repair Teams within the organic Brigade Support Battalion (BSB) to perform field maintenance requirements beyond the capabilities of the crew chief/crew, more in-depth Battle Damage Assessment Repair (BDAR), and limited recovery operations. The FRMV will carry a crew of three with additional space for two recovered crew members. The weapon system for the FRMV is the Close Combat Armament System (CCAS).

Unmanned Ground Vehicles

Armed Robotic Vehicle (ARV)

The Armed Robotic Vehicle (ARV) is an unmanned 9.5 ton 6x6 Hybrid Electric Drive (HED) Skid Steer vehicle and comes in two variants: the Assault variant and the Reconnaissance, Surveillance and Target Acquisition (RSTA) variant. The two variants share a common chassis. The Assault variant will support the mounted and dismounted forces in the assault with direct fire and anti-tank (AT) weapons providing LOS, BLOS targeting and over-watching fires; remotely occupies key terrain providing ISR/TA reconnaissance capability in MOUT and other battlespace; deploy sensors; locate or by-pass threat obstacles; assess battle damage, and acts as a communications relay. The RSTA version will remotely provide reconnaissance capability supporting Recon and MCS platoons in Urban Military Operations in Urban Terrain (MOUT) and other battlespace providing RSTA ISR/TA capability; deploy sensors, locate or by-pass threat obstacles; acts as a communications relay; and remotely assess and report battle damage assessment (BDA).

Small Unmanned Ground Vehicle (SUGV)

The Small Unmanned Ground Vehicle (SUGV) is a small, lightweight, manportable UGV capable of conducting military operations in urban terrain, tunnels, sewers and caves. The SUGV is an aid in enabling the performance of manpower intensive or high-risk functions (i.e. urban Intelligence, Surveillance, and Reconnaissance (ISR) missions, chemical/Toxic Industrial Chemicals (TIC)/Toxic Industrial Materials (TIM), reconnaissance, etc.) without exposing Soldiers directly to the hazard. The SUGV modular design allows multiple payloads to be integrated in a plug-and-play fashion. Weighing less than 30 pounds, it is capable of carrying up to six pounds of payload weight.

Multifunctional Utility/Logistics and Equipment (MULE) Vehicle

The Multifunctional Utility/Logistics and Equipment (MULE) Vehicle is a 2.5-ton Unmanned Ground Vehicle (UGV) that will support dismounted operations. It consists of four major components:

- Common Mobility platform (CMP)
- Three Mission Equipment Packages: Mule-Transport, ARV-A-L & Mule-Countermine
- Centralized Controller (CC) for Dismounted operations
- Autonomous Navigation System (ANS) mission payload package integrated on MULE platforms, Armed Robotic Vehicles and Manned Ground Vehicles (MGVs) to provide semiautonomous and leader-follower capability

The Multifunctional Utility/Logistics and Equipment (MULE) Vehicle is sling-loadable under military rotorcraft. The MULE Vehicle has three variants sharing a common chassis: transport, countermine and the Armed

Robotic Vehicle (ARV)-Assault-Light (ARV-A-L). The Transport MULE Vehicle (MULE-T) will carry 1,900-2,400 pounds of equipment and rucksacks for dismounted infantry squads with the mobility needed to follow squads in complex terrain. The Countermine MULE Vehicle (MULE-CM) will provide the capability to detect, mark and neutralize anti-tank mines by integrating a mine detection mission equipment package from the Ground Standoff Mine Detection System (GSTAMIDS) FCS (BCT) program. The ARV-Assault-Light (ARV-A-L) MULE Vehicle is a mobility platform with an integrated weapons and reconnaissance, surveillance and target acquisition (RSTA) package to support the dismounted infantry's efforts to locate and destroy enemy platforms and positions. The MULE Common Mobility Platform (CMP) is the program's centerpiece providing superior mobility built around the propulsion and articulated suspension system to negotiate complex terrain, obstacles and gaps that a dismounted squad will encounter.

Unmanned Aerial Vehicles

Class I Unmanned Aerial Vehicle (UAV)

The Class I Unmanned Aerial Vehicle (UAV) provides the dismounted soldier with Reconnaissance, Surveillance, and Target Acquisition (RSTA). Estimated to weigh less than 15 pounds, the air vehicle operates in complex urban and wooded terrains with a vertical take-off and landing capability. It is interoperable with selected ground and air platforms and controlled by dismounted soldiers. The Class I uses autonomous flight and navigation, but it will interact with the network and Soldier to dynamically update routes and target information. It provides dedicated reconnaissance support and early warning to the smallest echelons of the Brigade Combat Team (BCT) in environments not suited to larger assets. It will also perform limited communications relay in restricted terrain, a tremendous deficit in current operations.

The system (which includes two air vehicles, a control device, and ground support equipment) is back-packable.

Class II Unmanned Aerial Vehicle (UAV)

The Class II Unmanned Aerial Vehicle (UAV) has twice the endurance and a wider range of capabilities than the Class I. It is a multifunctional aerial system possessing the Vertical Take-Off and Landing capability. It supports the Company Commanders with reconnaissance, security/early warning, target acquisition and designation. The Class II Unmanned UAV will be a vehicle-mounted system that provides Line-of-Sight (LOS) enhanced dedicated imagery. The distinguishing capability of this UAV is target designation in day, night, and adverse weather. This provides the Company Commander the ability to shape the battlespace by employing a combination of Line-of-Sight (LOS), Beyond-Line-of-Sight (BLOS), and Non-Line-of-Sight (NLOS) fires. It can team with selected ground and air platforms, and provides limited communications relay.

The Class II Unmanned Aerial Vehicle (UAV) can be carried by two Soldiers.

Armed Robotic Test Vehicle
Summer 2006

Multifunctional Utility/Logistics and Equipment (MULE) test platform
Fall 2006

Small Unmanned Ground Vehicle (SUGV) demonstrating its capabilities at APG
Fall 2005

Class III Unmanned Aerial Vehicle (UAV)

The Class III Unmanned Aerial Vehicle (UAV) is a multifunction aerial system that has the range and endurance to support battalion level RSTA within the Brigade Combat Team's (BCT) battlespace. The Class III must maximize endurance and payload while minimizing maintenance, fuel, and transportation requirements. It provides the capabilities of the Class I and Class II, but also provides communications relay, mine detection, Chemical, Biological, Radiological, Nuclear, and High-yield Explosive (CBRNE) detection, and meteorological survey. It allows the Non-Line-of-Sight (NLOS) battalion to deliver precision fires within the BCT area of interest. It operates at survivable altitudes at standoff range during day, night and adverse weather. The Class III must be able to take-off and land without a dedicated air field.

Class IV Unmanned Aerial Vehicle (UAV)

The Class IV Unmanned Aerial Vehicle (UAV) has a range and endurance appropriate for the brigade mission. It supports the Brigade Combat Team (BCT) Commander with communications relay, long endurance persistent stare, and wide area surveillance. Unique missions include dedicated manned and unmanned teaming (MUM) with manned aviation; Wide Band Communications Relay; and standoff Chemical, Biological, Radiological, Nuclear, and High-yield Explosive (CBRNE) detection with on-board processing. Additionally, it has the payload to enhance the RSTA capability by cross-cueing multiple sensors. Like the Class III, the Class IV must be able to take-off and land without a dedicated air field.

Unattended Systems

Unattended Ground Sensors (UGS)

The FCS (BCT) Unattended Ground Sensors (UGS) program is divided into two major subgroups of sensing systems: Tactical-UGS (T-UGS), which includes Intelligence, Surveillance and Reconnaissance (ISR)-UGS and Chemical, Biological, Radiological and Nuclear (CBRN)-UGS; and Urban-UGS (U-UGS), also known as Urban Military Operations in Urban Terrain (MOUT) Advanced Sensor System (UMASS). The ISR-UGS will be modular and composed of tailorable sensor groups using multiple ground-sensing technologies. An Unattended Ground Sensors (UGS) field will include multimode sensors for target detection, location and classification; and an imaging capability for target identification. A sensor field will also include a gateway node to provide sensor fusion and a long-haul interoperable communications capability for transmitting target or SA information to a remote operator, or the common operating picture through the FCS (BCT) JTRS Network. The UGS can be used to perform mission tasks such as perimeter defense, surveillance, target acquisition and situational awareness (SA), including chemical, biological, radiological, nuclear, and high-yield explosive (CBRNE) early warning.

Urban-Unattended Ground Sensors (U-UGS) will provide a low cost, network-enabled reporting system for SA and force protection in an urban setting, as well as residual protection for cleared areas of Urban Military Operations in Urban Terrain (MOUT) environments. They can be hand-employed by Soldiers or robotic vehicles either inside or outside buildings and structures. U-UGS can support BCT operations by monitoring urban choke points such as corridors and stairwells as well as sewers, culverts and tunnels. U-UGS gateways provide the urban SA data interfaced to JTRS networks.

Non-Line of Sight-Launch System (NLOS-LS)

The Non-Line of Sight-Launch System (NLOS-LS) consists of a family of missiles and a highly deployable, platform-independent Container Launch Unit (CLU) with self-contained tactical fire control electronics and software for remote and unmanned operations. Each Container Launch Unit (CLU) will consist of a computer and communications system and 15 missiles [Precision Attack Missiles (PAM)].

Precision Attack Missiles (PAM) is a modular, multi-mission, guided missile with two trajectories—a direct-fire or fast-attack trajectory, and a boost-glide trajectory. The missile will receive target information prior to launch, and can receive and respond to target location updates during flight. The PAM will support laser-designated, laser-anointed and autonomous operation modes and will be capable of transmitting near-real-time information in the form of target imagery prior to impact. PAM is being designed to defeat heavy armored targets.

Intelligent Munitions System

The Intelligent Munitions System (IMS) is an unattended munitions system providing both offensive battlespace shaping and defensive force protection capabilities for the Future Force. The IMS is a system of lethal and non-lethal munitions integrated with robust command and control features, communications devices, sensors and seekers that make it an integral part of the FCS (BCT) network's core systems. IMS provides unmanned terrain dominance, economy of force and risk mitigation for the warfighting commander. Typical missions include:

- Isolating enemy forces, objectives, and areas of decisive operations.
- Creating lucrative targets, and engaging them or cueing other fires.
- Filling gaps in the noncontiguous battlespace.
- Controlling noncombatant movement with its non-lethal capabilities.

With its reduced footprint, IMS can be delivered by various means, and once on the ground, locate itself, organize all of its components and report its location to the Battle Command Mission Execution (BCME). It will be under positive control of the BCME, one of the FCS (BCT) command and control applications. The munition field can be armed, turned off to allow friendly passage, then rearmed to resume its mission. This on-off-on capability allows it to be recoverable, further reducing its logistics footprint. IMS will not become a residual hazard; it will self-destruct on command or at a preset time interval. It will also be tamper resistant.

Class 1 Micro Air Vehicle (MAV) demonstrating urban warfare capability during 25th ID ACTD testing
Fall 2006

Unmanned Ground Sensor (UGS) in demonstration at JEFX06
Summer 2006

Class IV Unmanned Air Vehicle demonstrator in flight at APG
Fall 2005

(18+1+1) The Network in Depth

The Army's FCS (BCT) network allows the FCS Family-of-Systems (FoS) to operate as a cohesive system-of-systems where the whole of its capabilities is greater than the sum of its parts. As the key to the Army's transformation, the network and its logistics and Embedded Training (ET) systems, enable the Army to employ revolutionary operational and organizational concepts. The network enables Soldiers to perceive, comprehend, shape, and dominate the future battlefield at unprecedented levels as defined by the FCS (BCT) Operational Requirements Document (ORD).

The FCS (BCT) network consists of five layers that when combined provides seamless delivery of data: the Standards, Transport, Services, Applications, and Sensors and Platforms Layers. The FCS (BCT) network possesses the adaptability and management functionality required to maintain pertinent services, while the FCS (BCT) fights on a rapidly shifting battlespace giving them the advantage to see first, understand first, act first, and finish decisively.

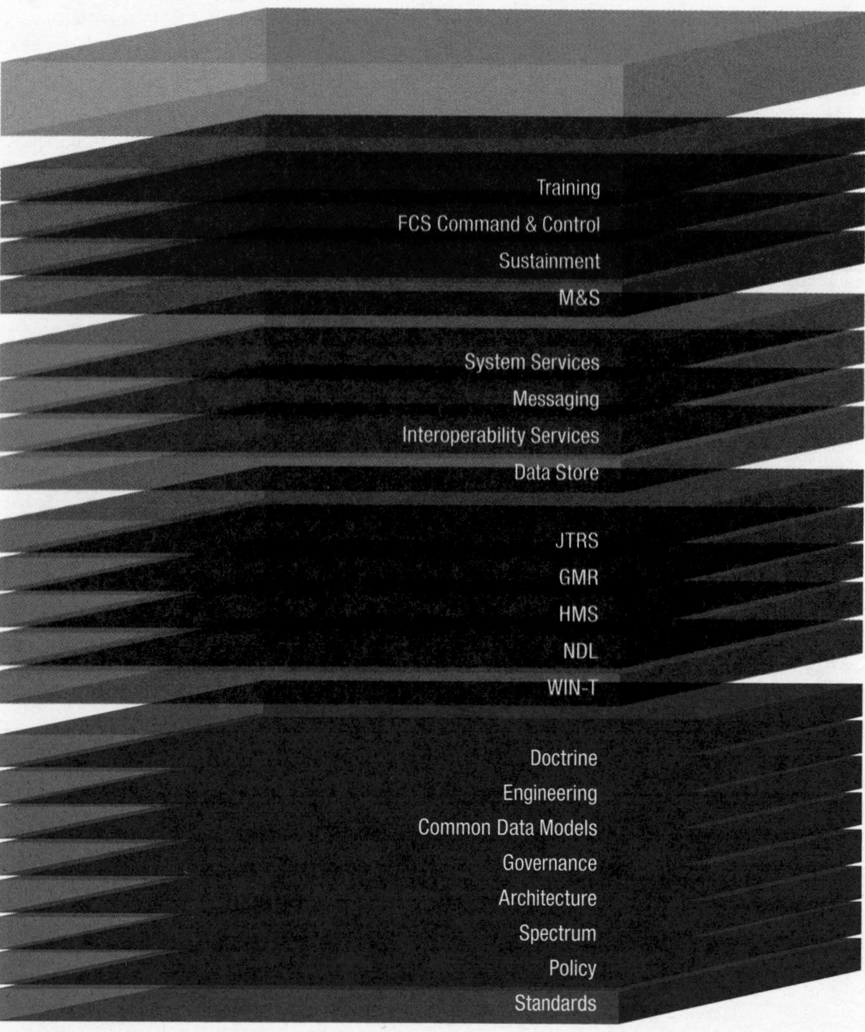

STANDARDS

The Standards Layer is the foundation of the FCS (BCT) network. It provides the governance for which the other layers are shaped and formed. The FCS (BCT) network will conform to the standards documentation to ensure that the net-centric attributes (flexible, adaptable distributed computing environment) are in place to move into the net-centric environment as part of a service-oriented architecture in the GIG. Information needs, information timeliness, information assurance, and netready attributes provide overarching guidance to ensure the technical exchange of information and the end-to-end operational effectiveness. Conformance to these standards permits seamless interoperability with combined and coalition forces for all National Security Systems (NSS) and Information Technology (IT) systems.

TRANSPORT

The FCS (BCT) Family-of-Systems (FoS) are connected to the command, control, communications, computers, intelligence, surveillance and reconnaissance (C4ISR) network by a multilayered transport layer with unprecedented range, capacity and dependability. The primarily mobile transport layer provides secure, reliable access to information sources over extended distances and complex terrain. The network will support advanced functionalities such as integrated network management, information assurance and information dissemination management to ensure dissemination of critical information among sensors, processors and warfighters both within, and external to the FCS (BCT)-equipped organization.

SERVICES

Central to FCS (BCT) network implementation is the Services Layer, commonly referred to as System-of-Systems Common Operating Environment (SOSCOE), which supports multiple mission-critical applications independently and simultaneously. It is configurable so that any specific instantiation can incorporate only the components that are needed for that instantiation. It enables straightforward integration of separate software packages, independent of their location, connectivity mechanism and the technology used to develop them.

APPLICATIONS

The Applications Layer is responsible for providing the integrated ability to assess, plan, and execute network-centric mission operations using a common interface and a set of non-overlapping functional services that provides the full range of FCS (BCT) Warfighter capabilities.

The Applications Layer combines ten software packages to enable full interaction, integration, and interoperability between systems with no hardware, software or information stovepipes. It also allows cross Battlefield Functional Area (BFA) problem-solving; decision aiding; adaptable doctrine, tactics, techniques, and procedures; reconfiguration of roles and levels of automation during execution; development and efficiencies promotion; and technology refresh and insertion.

PLATFORM & SENSORS

The Sensors and Platforms Layer is comprised of a distributed and networked array of multi-spectral sensors that provide the FCS (BCT) with the ability to "see first." Intelligence, Surveillance and Reconnaissance (ISR) sensors will be integrated onto all manned ground vehicles, all unmanned ground vehicles and all four classes of unmanned aerial vehicles within the FCS (BCT).

To provide warfighters with current, accurate, and actionable information, the data from the various distributed ISR and other external sensor assets are subject to complex data processing, filtering, correlation, and aided target recognition and fusion.

The 18 networked systems consist of eight manned ground vehicles, three unmanned ground vehicles, four unmanned aerial vehicles, and three specialized devices.

(18+1+1) The Soldier—The Heart of FCS

All Soldiers in the Brigade Combat Team (BCT) are part of the Soldier as a System (SaaS) overarching requirement that encompasses everything the Soldier wears, carries, and consumes to include unit radios, crew served weapons, and unit specific equipment in the execution of tasks and duties.

All Soldiers systems will be treated as an integrated System of Systems (SoS). The Soldier, as defined by Soldier as a System (SaaS), meets the need to improve the current capability of all Soldiers, regardless of Military Occupational Specialty (MOS), to perform Army Warrior Tasks and functions more efficiently and effectively. Soldier as a System (SaaS) establishes a baseline for core Soldier requirements, and establishes the foundation for specific or mission unique Soldier Programs (Ground, Mounted, and Air). It will present a fully integrated modular Soldier that provides a balance of tasks, and mission equipment in support of the Soldier Team, the Current, and the Future Force.

FCS also enhances the SaaS with additional benefits like joint embedded training—allowing the Soldier to train anywhere, at any time, including enroute to the battlefield.

On cost, on schedule, and performing

FCS is transformational. The Army is accelerating the fielding of individual FCS systems and technologies to the Current Force as they become available. These "spin outs" will provide real-time benefits to our Soldiers who may be in harm's way. The Army's plan is to ensure that all elements of the Current Force will benefit from FCS technology improvements by providing them with FCS capability sooner rather than later.

FY06 was a critical year for FCS, with 52 major reviews, broad industry ramp up, progress on the network and platforms, and extensive hardware and software deliveries. The program successfully conducted its first major field experiments (Experiment 1.1, JEFX06) and is now positioned to deliver initial spin out technologies in FY08.

During FY07, the FCS program will conduct research designed to bring FCS capabilities to the Current Force. Specifically, the program is looking at improvements to the Abrams, Bradley, and HMMWV systems to develop precision networked fires, enhance force protection, and improve interoperability.

In addition, features of FCS systems, such as unattended ground sensors (UGS), the non-line-of sight systems, and intelligent munitions, will be made ready for Spin Out 1. Also during FY07, the program will conduct network testing and verification, develop software builds, and build prototypes of FCS systems.

Spin Out 1 is currently under development. Program acquisition controls are in place, and all systems within Spin Out 1 are progressing through key engineering milestones. Spin Out 1 will begin issuing in Fiscal Year (FY) 2008 and consist of prototypes issued to the Evaluation Brigade Combat Team (EBCT) for their use and evaluation. Following successful evaluation, production and fielding of Spin Out 1 will commence to Current Force units in 2010. This process will be repeated for each successive Spin Out. By 2014, the Army force structure will include one Brigade Combat Team (BCT) equipped with all 18 + 1 FCS (BCT) core systems and additional Brigade Combat Teams with embedded FCS (BCT) capability.

Key Milestones

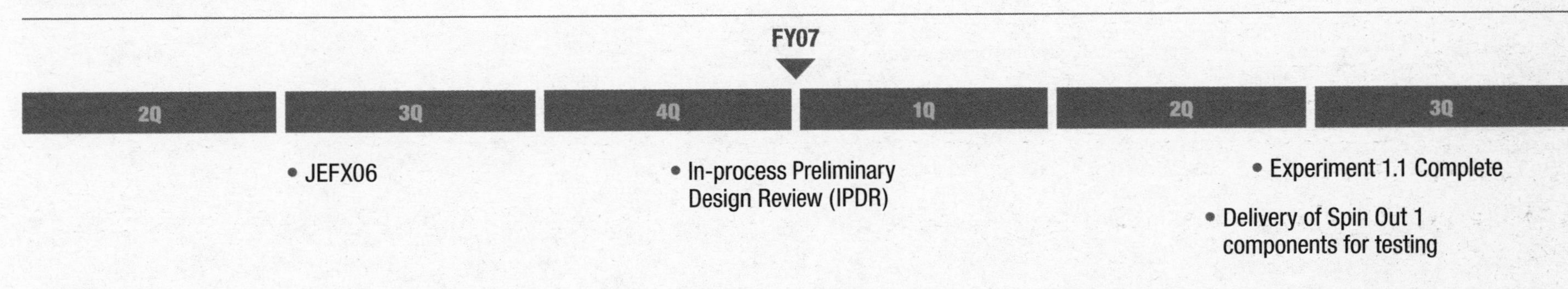

Weapon Systems

Listed in alphabetical Order

2.75" Family of Rockets

Provides affordable air-to-ground precision fires to defeat soft and lightly armored targets (Advanced Precision Kill Weapon System [APKWS]) and suppression, illumination, and direct/indirect fires to defeat area, materiel, and personnel targets (Hydra 70).

INVESTMENT COMPONENT

- Modernization
- Recapitalization
- Maintenance

Description & Specifications

The Advanced Precision Kill Weapon System (APKWS) and Hydra 70 Rocket System make up the family of 2.75" air-launched rockets employed by tri-service and special operating forces on both fixed wing and rotary wing aircraft. This highly modular rocket family incorporates a laser guidance section mated to a high-explosive warhead for the APKWS variant, and several different mission-oriented warheads for the Hydra 70 variant, including high-explosive, multi-purpose submunition, red phosphorus smoke, flechette, visible light illumination flare, and infrared illumination flare.

Program Status

APKWS

- **3QFY05:** Program restructure decision
- **2QFY06:** Milestone decision authority restructure in process review
- **3QFY06:** Contract awarded to continue system development and demonstration activities

Hydra 70

- **Current:** Producing annual replenishment requirements for training and war reserve

Recent and Projected Activities

APKWS

- **4QFY08:** Low rate initial production (LRIP) (Milestone C) decision review
- **1QFY09:** Initiate LRIP

Hydra 70

- **Ongoing:** Continue Hydra 70 production and insensitive munitions improvement activities

ACQUISITION PHASE

Concept & Technology Development | System Development & Demonstration | Production & Deployment | Operations & Support

2.75" Family of Rockets

DIAMETER: 2.75 inches
WEIGHT: APKWS, 35 pounds; Hydra 70 23–27 pounds (depending on warhead)
LENGTH: APKWS, 75 inches; Hydra 70 55–70 inches (depending on warhead)
RANGE: APKWS, 1,500–5,000 meters; Hydra 70, 300–8,000 meters

FOREIGN MILITARY SALES

APKWS: None
Hydra 70: Kuwait, the Netherlands, Colombia, Singapore, Thailand, United Arab Emirates, and Japan

CONTRACTORS

APKWS:
BAE Systems Electronics & Integrated Solutions (Nashua, NH)
Hydra 70:
General Dynamics Armament and Technical Products (GDATP) (Burlington, VT)
Grain:
Alliant Techsystems (Radford, VA)
Fuzes:
Action Manufacturing (Philadelphia, PA)
Fin and nozzle:
General Dynamics Ordnance and Tactical Systems (Anniston, AL)

Abrams Upgrade

Provides mobile, protected firepower for battlefield superiority.

INVESTMENT COMPONENT

Modernization

Recapitalization

Maintenance

Description & Specifications

The Abrams upgrade includes two variants, the M1A1 and M1A2, and provides the lethality, survivability, and fightability to defeat advanced threats on the integrated battlefield using mobility, firepower, and shock effect. The 120mm main gun on the M1A1 and M1A2 and the 1,500-horsepower AGT turbine engine and special armor make the Abrams tank particularly lethal against heavy armor forces.

The M1A1 upgrade includes increased armor protection, suspension improvements, and a nuclear, biological, and chemical (NBC) protection system for survivability. An integrated applique computer, an embedded diagnostic system, a second-generation thermal sensor, and a far-target-designation capability can be incorporated on the M1A1.

The M1A2 upgrade includes a commander's independent thermal viewer, an improved weapon station, position navigation equipment, distributed data and power architecture, embedded diagnostics, and an improved fire control system.

The M1A2 system enhancement program (SEP) adds second-generation thermal sensors and a thermal management system. It includes upgrades to processors/memory that enable use of the Army's common command and control software, effecting rapid transfer of digital situational data and overlays.

The Abrams upgrade includes the total integrated revitalization (TIGER) program, a rebuild effort consisting of engine data collection, transition of parts management to the contractor, and implementation of commercial production processes. The Abrams integrated management (AIM) overhaul program recapitalizes the high operational tempo of the M1A1 tank fleet. The Abrams parts obsolescence program tracks obsolete components for spare parts and maintains a database of current parts.

Program Status

- **FY07:** M1A2 SEP upgrade production complete (547 total)
- **FY07:** M1A2 to M1A2 SEP retrofit program continues to equip the 3rd Armored Cavalry Regiment (ACR)
- **FY07:** M1A1 AIM continues fielding to the 3rd Infantry Division (ID)

Recent and Projected Activities

- **FY07–09:** M1A1 AIM continues fielding to the Army National Guard (ARNG), 1st ID, 2nd ID, and TRADOC
- **FY07–09:** M1A2 SEP production
- **FY07–09:** Continue fielding M1A2 SEP to the 3rd ACR, 1st Cavalry Division, and 1st Armored Division (AD), and M1A1 AIM to the ARNG, 1st AD, and 2nd ID
- **FY07:** TIGER production begins with initial deliveries beginning May 2007

ACQUISITION PHASE

Concept & Technology Development | System Development & Demonstration | Production & Deployment | Operations & Support

Abrams Upgrade

FOREIGN MILITARY SALES
M1A1: Australia (59); Egypt (1,005)
M1A2: Kuwait (218); Saudi Arabia (315)

CONTRACTORS
General Dynamics (Sterling Heights, MI; Warren, MI; Muskegon, MI; Scranton, PA; Lima, OH; Tallahassee, FL)
Honeywell (Phoenix, AZ)
Raytheon (McKinney, TX)

	M1/1PM1	M1A1	M1A2	M1A2 SEP
LENGTH (feet):	32.04	32.04	32.04	32.04
WIDTH (feet):	12.0	12.0	12.0	12.0
HEIGHT (feet):	7.79	8.0	8.0	8.0
TOP SPEED (mph):	45.0	41.5	41.5	42
WEIGHT (tons):	61.4/62.8	67.6	68.4	69.5
ARMAMENT:	105mm	120mm	120mm	120mm
CREW:	4	4	4	4

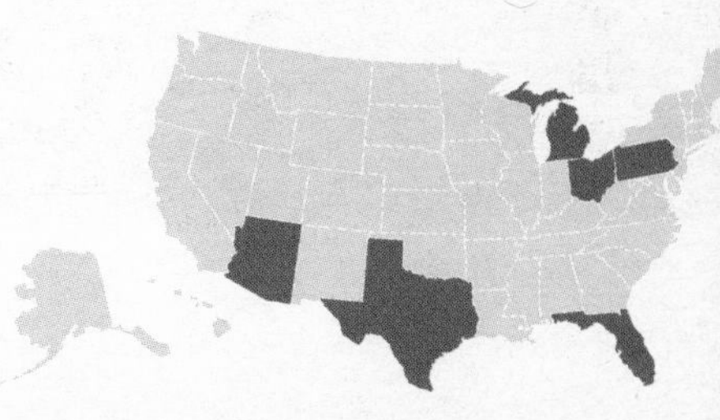

Advanced Field Artillery Tactical Data System (AFATDS)

Provides automated fire support command, control, and communications for the Army, Navy, and Marine Corps.

INVESTMENT COMPONENT

Modernization

Recapitalization

Maintenance

Description & Specifications

The Advanced Field Artillery Tactical Data System (AFATDS) performs the attack analysis necessary to determine optimal weapon-target pairing to provide automated planning, coordination, and control for maximum use of fire support assets (field artillery, mortars, close air support, naval gunfire, attack helicopters, and offensive electronic warfare).

AFATDS performs the fire support command, control, and coordination requirements of field artillery and maneuver from echelons above corps to battery or platoon in support of all levels of conflict. The system is composed of common hardware and software employed in varying configurations at different operational facilities (or nodes) and unique system software interconnected by tactical communications in the form of a software-driven, automated network.

AFATDS will automatically implement detailed commander's guidance in the automation of operational planning, movement control, targeting, target value analysis, and fire support planning. This project is a replacement system for the Initial Fire Support Automated System, Battery Computer System, and Fire Direction System. AFATDS is designed to interoperate with the other Army battle command systems; current and future Navy and Air Force command and control weapon systems; and the German, French, British, and Italian fire support systems.

Program Status

- **1QFY05–4QFY05:** Army operational assessment software blocking (SWB) (6.4)
- **1QFY06:** Urgent materiel release of AFATDS 6.4 to Operation Iraqi Freedom and Operation Enduring Freedom

Recent and Projected Activities

- **3QFY06–2QFY07:** AFATDS 6.5 test fix test cycle (SWB 08-10)
- **2QFY07:** AFATDS 6.4 full materiel release and fielding
- **3QFY07:** AFATDS 6.5 Intra-Army interoperability certification
- **2QFY08:** AFATDS 6.5 full materiel release

ACQUISITION PHASE

Concept & Technology Development | System Development & Demonstration | Production & Deployment | Operations & Support

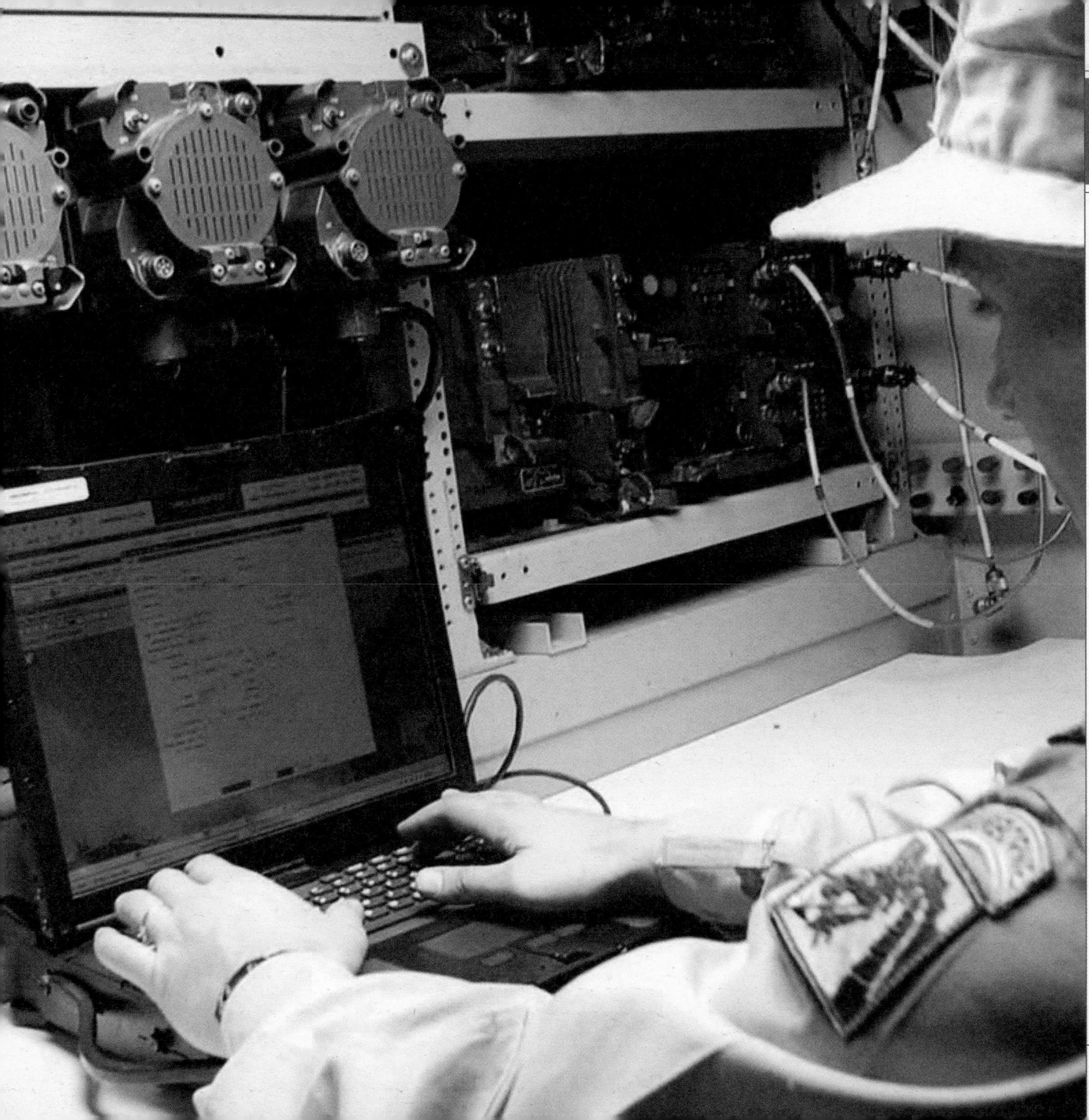

Advanced Field Artillery Tactical Data System (AFATDS)

FOREIGN MILITARY SALES
Bahrain, Egypt, Portugal, Turkey

CONTRACTORS
Software:
Raytheon (Ft. Wayne, IN)
Hardware:
General Dynamics (Taunton, MA)
Technical support:
Computer Sciences Corp. (Tinton Falls, NJ)
New equipment training:
Engineering Professional Services (Lawton, OK)
Testing:
Titan Corp. (Lawton, OK)

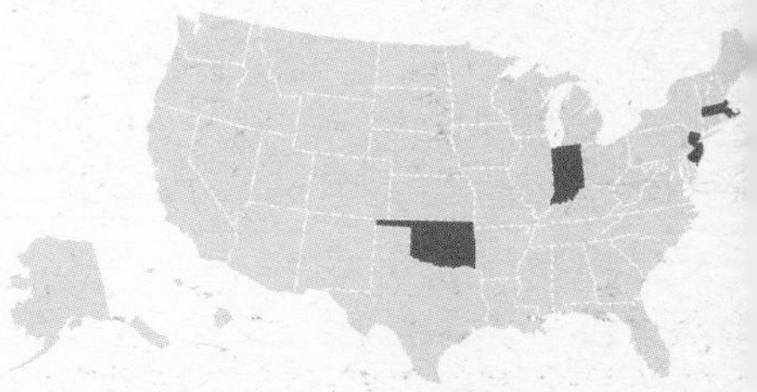

Advanced Threat Infrared Countermeasures (ATIRCM)/ Common Missile Warning System (CMWS)

Provides an airborne countermeasure system that detects missile launches/flights and protects aircraft against infrared missiles.

INVESTMENT COMPONENT

Modernization

Recapitalization

Maintenance

Description & Specifications

The Advanced Threat Infrared Countermeasures (ATIRCM)/ Common Missile Warning System (CMWS) integrates defensive infrared countermeasures capabilities into existing, current generation aircraft for more effective protection against a greater number of guided missile threats than is afforded by currently fielded infrared countermeasures. The U.S. Army operational requirements concept for infrared countermeasure systems is the Suite of Integrated Infrared Countermeasures (SIIRCM). It mandates an integrated warning and countermeasure system to enhance aircraft survivability against infrared guided threat missile systems. The core element of the SIIRCM concept is the ATIRCM/CMWS Program. The ATIRCM/CMWS, a subsystem to a host aircraft, consists of an integrated ultraviolet missile warning system, an Infrared Laser Jammer, and Improved Countermeasure Dispensers (ICMDs).

The CMWS can function as a stand-alone system with the capability to detect missiles and provide audible and visual warnings to pilots. When installed with the ICMD, it activates expendables to decoy/defeat infrared-guided missiles. ATIRCM adds infrared laser jamming to CMWS, and is a key survivability system for Future Force Army aircraft.

The A-Kit contains the modification hardware, wiring harness, cable, and other items required to install and interface the ATIRCM/CMWS Mission Kit to each platform. The A-Kit ensures the Mission Kit is functionally and physically operational with specific host platforms.

The Mission Kit, also known as the B-Kit, consists of the ATIRCM/ CMWS components that perform the missile detection, false alarm rejection, missile declaration, and countermeasure functions of the system. The CMWS Electronic Control Unit receives ultraviolet missile detection data from electro-optic missile sensors and sends a missile alert signal to on-board avionics and ATIRCM. Threat missiles detected by the CMWS are subsequently tracked and jammed by ATIRCM and ICMD. Working together, CMWS/ICMD and ATIRCM provide a combination of missile seeker and countermeasures including decoy flares and laser jamming to defeat all known Tier 1 threat missiles.

Program Status

- **3QFY06:** CMWS full rate production

Recent and Projected Activities

- **Ongoing through 1QFY09:** ATIRCM reliability demonstration testing

ACQUISITION PHASE

Concept & Technology Development | System Development & Demonstration | Production & Deployment | Operations & Support

Advanced Threat Infrared Countermeasures (ATIRCM)/Common Missile Warning System (CMWS)

SYSTEM WEIGHT (includes A-Kit weight):

ATIRCM/CMWS: Ranges from 189–313 pounds, depending on platform type
CMWS and improved countermeasure dispenser only: Ranges from 80–150 pounds, depending on platform type

FOREIGN MILITARY SALES
None

CONTRACTORS
BAE Systems (Nashua, NH)
L-3 Communications (Lexington, KY)
DynCorp International (Ft. Worth, TX)
Westwind Technologies, Inc. (Huntsville, AL)
Sikorsky Aircraft Corp. (Stratford, CT)

Aerial Common Sensor (ACS)

Provides global, real-time multi-intelligence precision targeting information to joint land, maritime, and air combat commanders across the full spectrum of military operations.

INVESTMENT COMPONENT

Modernization

Recapitalization

Maintenance

Description & Specifications

The Aerial Common Sensor (ACS) delivers the intelligence, surveillance, and reconnaissance (ISR) data required for dominant maneuver, precision engagement, and decision superiority by merging and enhancing the sensor capabilities of current ISR platforms, Airborne Reconnaissance Low (ARL), and Guardrail Common Sensor (GR/CS). ACS provides a larger area of coverage, supporting Future Force operational geometries and the greater lethality ranges of new weapons systems.

ACS transforms Army airborne ISR from a strategic-lift-intensive, maximum-deployment-time asset to a minimum-lift, minimal-deployment-time, global asset capable of operation immediately upon arrival into theater. Mission tailorable and scalable, ACS provides distributed, wide area, persistent surveillance and multi-intelligence precision targeting responsive to the ground tactical commander.

Using robust sensor-to-shooter and reach-back links, ACS provides multi-sensor intelligence throughout a non-linear framework and non-contiguous battlespace providing real-time sensor-to-shooter information. The system will be fully interoperable with joint and national collectors, ground processing facilities, and dissemination systems, meeting transformational, joint net-centric situational awareness requirements, as well as Army requirements for a worldwide, self-deployable single ISR system.

ACS supports warfighter requirements across the full spectrum of operations, from early/denied entry through crisis resolution, including critical precision signal intelligence (SIGINT) linkage into the joint ISR Network and Imagery intelligence (IMINT).

ACS will replace current airborne ISR systems in all five of the Army's Aerial Exploitation Battalions.

Program Status

- **2QFY06:** ACS system development and demonstration contract terminated for convenience of the government
- **4QFY06:** Program assessments via joint ISR study results and requirement validation
- **1QFY07:** Requirements determination

Recent and Projected Activities

- **2QFY07:** Acquisition strategy development

ACQUISITION PHASE

Concept & Technology Development | System Development & Demonstration | Production & Deployment | Operations & Support

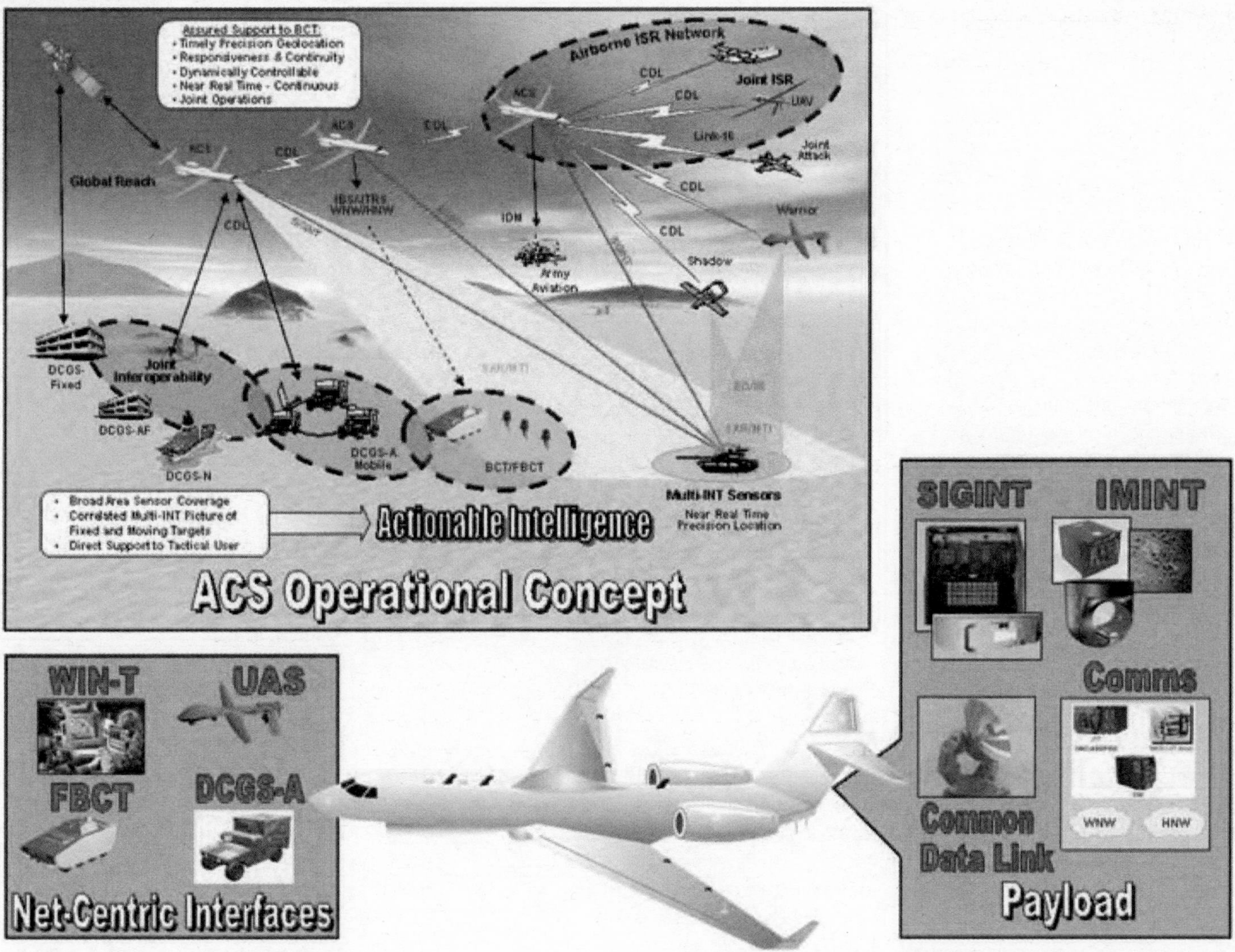

Aerial Common Sensor (ACS)

FOREIGN MILITARY SALES
None

CONTRACTORS
To be determined

Air Warrior (AW)

Enhances Army aircrew comfort, cockpit synergy, and aircraft mission capability, and improves Soldier lethality, survivability, mobility, and sustainability through a new generation, integrated aircrew ensemble.

INVESTMENT COMPONENT

- Modernization
- Recapitalization
- Maintenance

Description & Specifications

Air Warrior (AW) provides advanced life support, ballistic protection, and chemical/ biological (CB) protection in a single system comprising rapidly tailorable, mission-configurable modules. Air Warrior addresses interoperability and has leveraged several joint-service technology efforts to effectively integrate all functions into a single system. The system maximizes safe aircraft operation and human performance while not encumbering the aircrew. Components include:

- Microclimate cooling system
- Digital connectivity and situational awareness
- Wireless intercom
- Aircraft platform interface
- CB protection
- Body armor
- Survival and escape and evasion tools
- Overwater survival items

Improvements to the Air Warrior system are incrementally provided through an evolutionary acquisition program to solve equipment shortcomings. Components include the following:

- Block 2 Electronic Data Manager (EDM) is a miniature, knee-held computer that provides two-way situational awareness—Blue Force Tracking plus GPS-based moving map—and digital mission planning capability to current aircraft fleet.
- Air Warrior Modular Integrated Helmet Display System (MIHDS) will be a Block 3 advanced capabilities program.

The Air Warrior system is the key ingredient to closing the performance gap between the aircrew and the aircraft. Air Warrior is answering the aviation warfighter challenges of today and tomorrow by developing affordable, responsive, deployable, versatile, lethal, survivable, and sustainable aircrew equipment.

Program Status

- **1QFY06:** Began full qualification testing of the Spiral 2 EDM hardware and software
- **2QFY06:** Production decision for the Spiral 2 EDM
- **2QFY06:** Began development of the Spiral 3 EDM
- **4QFY06:** Awarded encryption contract for Aircraft Wireless Intercom System (AWIS) program
- **1QFY07:** Block 1 and EDM fielding continues to deploying Operation Iraqi Freedom and Operation Enduring Freedom units

Recent and Projected Activities

- **4QFY07:** Qualify and begin fielding the Spiral 3 EDM
- **2QFY08:** Complete development of the encrypted AWIS
- **1QFY09:** Complete fielding of the EDM kit on AH-64A/D, OH-58D, and CH-47D aircraft
- **2QFY09:** Full rate production decision review for encrypted AWIS

ACQUISITION PHASE

- Concept & Technology Development
- System Development & Demonstration
- Production & Deployment
- Operations & Support

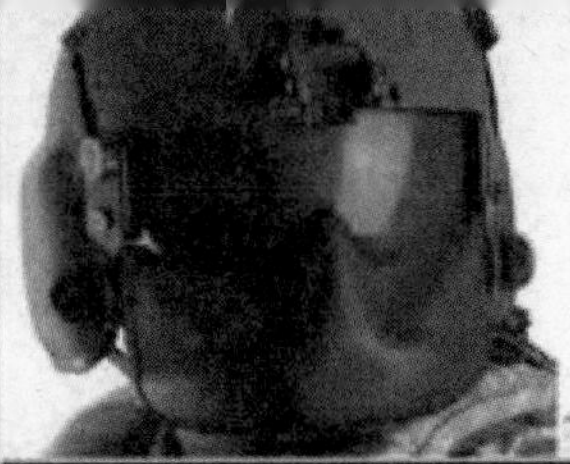

Air Warrior (AW)

FOREIGN MILITARY SALES
Australia

CONTRACTORS
Carleton Technologies, Inc. (Orchard Park, NY)
Simula Aerospace and Defense Group, Inc. (Phoenix, AZ)
Aerial Machine and Tool, Inc. (Vesta, VA)
Armor Holdings, Inc. (Ontario, CA)
Westwind Technologies, Inc. (Huntsville, AL)
Raytheon Technical Services, Inc. (Indianapolis, IN)
Secure Communications Systems, Inc. (Santa Ana, CA)
Telephonics Corp. (Farmingdale, NY)
General Dynamics C4 Systems, Inc. (Scottsdale, AZ)

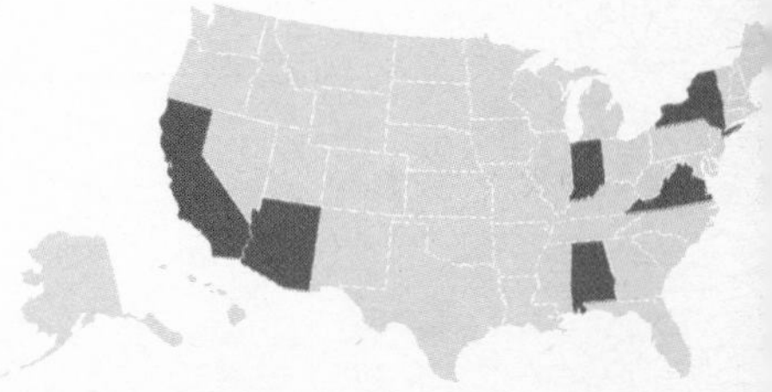

Air/Missile Defense Planning and Control System (AMDPCS)

Provides an automated command and control system that enables commanders to see the battlespace and plan for attack and controls artillery engagements through a single, integrated system.

INVESTMENT COMPONENT

- **Modernization**
- Recapitalization
- Maintenance

Description & Specifications

The Air and Missile Defense Planning and Control System (AMDPCS) provides integration of Air and Missile Defense operations at all echelons. It is a planning and battlespace situational awareness tool that provides commanders with a common tactical and operational air picture. AMDPCS systems are deployed with Air Defense Artillery Brigades, Army Air Missile Defense Commands (AAMDCs), Air Defense Airspace Management (ADAM) cells at Brigade Combat Teams (BCTs), fire brigades, and divisions.

Essential in filling the Army's modularity requirement, ADAM cells provide the commander at BCTs, brigades, and divisions with air defense situational awareness and airspace management capabilities. They also provide the interoperability link with joint, multi-national, coalition forces.

AMDPCS provides these organizations with shelters, automated data processing equipment, tactical communications, standard vehicles, tactical power, and the two major software systems used in air defense force operations/engagement operations: Air and Missile Defense Workstation (AMDWS) and Air Defense System Integrator (ADSI). ADSI monitors and controls air battle engagement operations by subordinate or attached units.

Program Status

- **3QFY06:** Army Battle Command test
- **4QFY06:** 40 ADAMs deployed; counter-rockets, artillery, mortars demonstrated
- **1QFY07:** ADAM cell Milestone C full rate production

Recent and Projected Activities

- **4QFY07:** Complete fielding 23 ADAMs; 32nd AAMDC fielding
- **4QFY08:** Complete fielding 28 ADAMs
- **1QFY09:** Begin fielding eight ADAMs

ACQUISITION PHASE

Concept & Technology Development | System Development & Demonstration | **Production & Deployment** | Operations & Support

Air/Missile Defense Planning and Control System (AMDPCS)

FOREIGN MILITARY SALES
None

CONTRACTORS
Sheltered systems and AMDWS software:
Northrop Grumman (Huntsville, AL)
ADSI software and hardware:
APC (Austin, TX)

Airborne Reconnaissance Low (ARL)

Detects, locates, and reports threat activities using a variety of imagery, communications-intercept, and moving-target indicator sensor payloads.

INVESTMENT COMPONENT

Modernization

Recapitalization

Maintenance

Description & Specifications

Airborne Reconnaissance Low (ARL) is a self-deploying, multi-sensor, day/night, all-weather reconnaissance, intelligence, echelons-above-corps asset. It consists of a modified DeHavilland DHC-7 fixed-wing aircraft equipped with communications intelligence (COMINT), imagery intelligence (IMINT), and synthetic aperture radar/moving target indicator (SAR/MTI) mission payloads. The payloads are controlled and operated via on-board open-architecture, multi-function workstations.

Intelligence collected on the ARL can be analyzed, recorded, and disseminated on the aircraft workstations in real time and/or stored on board for post-mission processing. During multi-aircraft missions, data can be shared between cooperating aircraft via ultra high frequency (UHF) air-to-air data links allowing multi-platform COMINT geolocation operations. The ARL system includes a variety of communications subsystems to support near-real-time dissemination of intelligence and dynamic retasking of the aircraft.

There are currently two configurations of the ARL system:

- Two aircraft are configured as ARL-COMINT (ARL-C), with a conventional communications intercept and direction finding (location) payload.
- Six aircraft are configured as ARL-Multifunction (ARL-M), equipped with a combination of IMINT, COMINT, and SAR/MTI payload and demonstrated hyperspectral imager applications and multi-INT data fusion capabilities.

Southern Command (SOUTHCOM) operates two ARL-C and three ARL-M aircraft. United States Forces Korea (USFK) operates three ARL-M.

Planned upgrades for ARL include baselining the fleet by providing a common architecture for sensor management and workstation Man-Machine Interface (MMI). ARL-C systems will be converted from COMINT only to ARL-M Multi-INT configuration. Planned sensor improvements include upgrading the radar to provide change detection and super resolution SAR, upgrading the MX-20 EO/IR subsystem to reflect current standards, including the addition of a laser illuminator, and the addition of Digital Pan cameras across the fleet for high-resolution imaging and change detection. A new and improved COMINT payload will be fielded, increasing frequency coverage and improving target intercept probability.

Program Status

- **4QFY05:** Federated COMINT upgrades on M4, M5, and M6 fielded
- **1QFY07:** HISAR radar replacement on M1, M2, and M3 (Korea completed)

Recent and Projected Activities

- **3QFY07:** Begin cockpit and workstation architecture standardization across fleet
- **4QFY07:** Standardize MX-20 video sensors across fleet
- **2QFY08:** Complete Phoenix Eye upgrade on long-range Ground Moving Target Indicator/Synthetic Aperture Radar (GMTI/SAR)
- **2QFY08:** Convert ARL C1 and C2 into ARL M7 and M8

ACQUISITION PHASE

Concept & Technology Development | System Development & Demonstration | Production & Deployment | Operations & Support

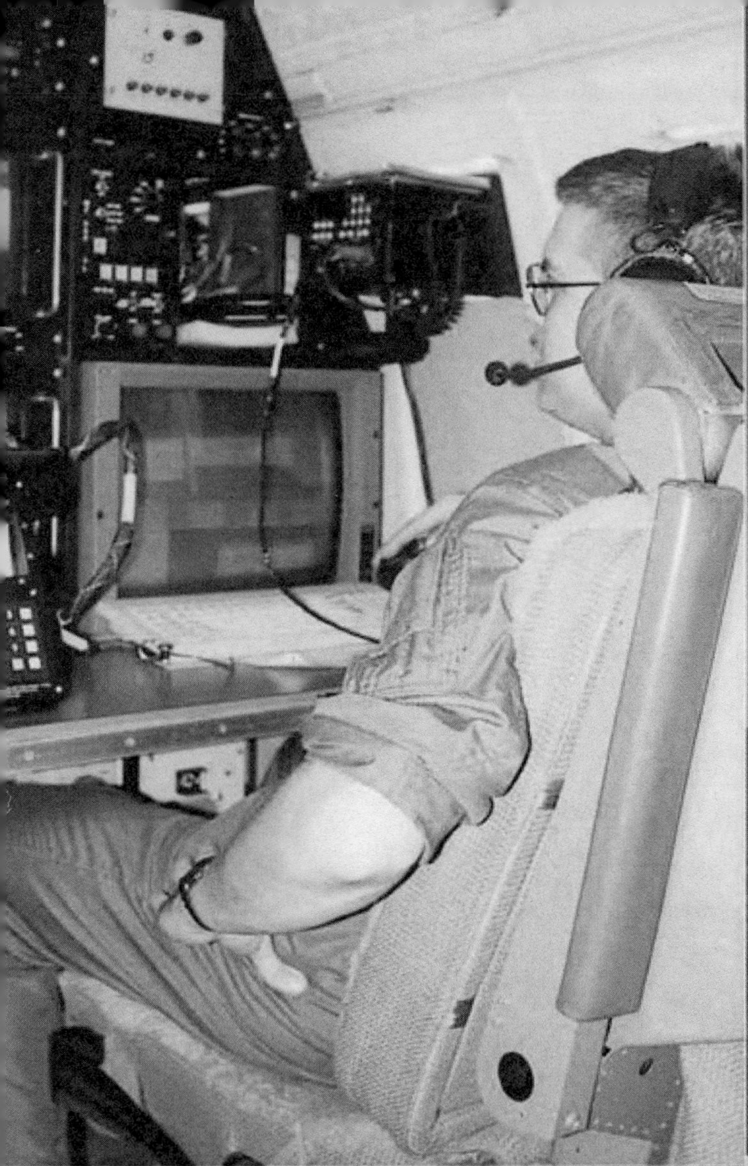

Airborne Reconnaissance Low (ARL)

FOREIGN MILITARY SALES
None

CONTRACTORS
Aircraft modifications:
Sierra Nevada Corp. (Hagerstown, MD)
Aircraft survivability:
Litton Advanced Systems (Gaithersburg, MD)
COMINT subsystem:
BAE Systems (Manchester, NH)
EO/IR subsystem:
WESCAM (Hamilton, Ontario, Canada)
Engineering support:
CACI (Berryville, VA)
Radar subsystem:
Lockheed Martin (Phoenix, AZ)

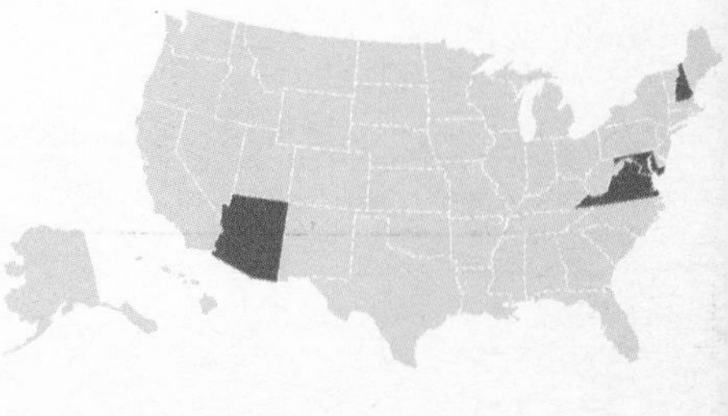

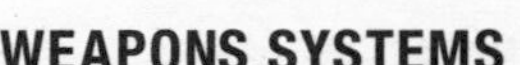

All Source Analysis System (ASAS)

Provides automated support to intelligence analysis, manages intelligence and electronic warfare resources, and produces and disseminates intelligence to commanders and staff.

INVESTMENT COMPONENT

Modernization

Recapitalization

Maintenance

Description & Specifications

The All Source Analysis System (ASAS) receives and correlates information from national, strategic, and tactical intelligence sensors and sources. It produces a correlated ground picture; disseminates intelligence products; provides target nominations; manages ISR collection and mission planning; and provides counterintelligence and electronic warfare. ASAS supports current operations and future planning at all echelons during all phases of military operations, across the full spectrum of conflict.

Block II ASAS is built on tactically deployable, ruggedized, commercial laptop and desktop computers. Block II software lets units track and link individuals, events, and organizations in stability and security operations, a critical capability for supporting the Global War on Terror and Operations Iraqi Freedom and Enduring Freedom. The Block II ASAS Analysis and Control Element (ACE) performs all source fusion, combining open sources with Signals Intelligence, Human Intelligence, Imagery Intelligence, and Measurement and Signature Intelligence. This Level 1 intelligence fusion capability will migrate into the Distributed Common Ground Station-Army (DCGS-A) in FY08.

Block II ASAS includes:

- The ACE, found at division, corps, and echelon above corps
- The ASAS-Light (ASAS-L) laptop configuration and ASAS-L Intelligence Fusion Station, [ASAS-L(IFS)], which supports S2s and intelligence elements from battalion to Army Service Component Commands
- The Analysis and Control Team-Element (ACT-E), which is shelter-mounted on a HMMWV and supports Brigade Combat Team (BCT) S2s

Program Status

- **2QFY05–1QFY07:** Provided interim intelligence fusion to DCGS-A
- **2QFY05–1QFY07:** Fielding and training of ASAS-L, IFS, and ACT-E with priority to deploying units
- **3QFY05:** Conducted Block II ASAS ACE operational test and Milestone C review
- **1QFY06:** Fielded Block II ACE to first unit equipped, 4th Infantry Division
- **2QFY06:** Fielded Block II ACE to 1st Cavalry Division
- **3QFY06:** Fielded Block II ACE to 25th Infantry Division and 82nd Airborne Corps
- **4QFY06:** Fielded Block II ACE to 513th Military Intelligence Brigade

Recent and Projected Activities

- **2QFY07–1QFY09:** Migrate ASAS capabilities into DCGS-A
- **2QFY07–2QFY09:** Fielding and training of DCGS-A capable ASAS-Light, IFS, ACE, and ACT-E with priority to deploying units

ACQUISITION PHASE

Concept & Technology Development | System Development & Demonstration | Production & Deployment | Operations & Support

All Source Analysis System (ASAS)

FOREIGN MILITARY SALES
None

CONTRACTORS
Hardware:
General Dynamics (Taunton, MA)
SETA support:
Lockheed Martin Integrated Systems, Inc. (Wall, NJ)
Fielding/maintenance support:
MANTECH (Killeen, TX)

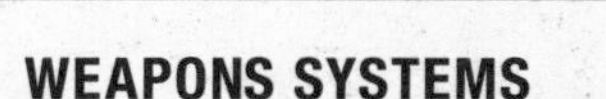

Analytical Laboratory System-System Enhancement Program (ALS-SEP)

Enables National Guard weapons of mass destruction-Civil Support Teams to perform on-site analysis of unknown samples in support of first responders with a mobile laboratory.

INVESTMENT COMPONENT

Modernization

Recapitalization

Maintenance

Description & Specifications

The Analytical Laboratory System (ALS) Increment 1 is a mobile analytical laboratory that provides the Civil Support Team (CST) capabilities for detecting and identifying chemical, biological, or radiological contamination.

ALS Increment 1 is a system enhancement program (SEP) to replace the current Mobile Analytical Laboratory System and interim Dismounted Analytical Platform. It provides advanced technologies with enhanced sensitivity and selectivity in the detection and identification of biological and chemical warfare agents and toxic industrial chemicals and materials.

Program Status

- **4QFY06:** Systems integration decision review

Recent and Projected Activities

- **3QFY07:** Start operational test
- **1QFY08:** Upgrade and deployment in progress review
- **2QFY08:** First unit equipped

ACQUISITION PHASE

Concept & Technology Development | System Development & Demonstration | Production & Deployment | Operations & Support

Analytical Laboratory System-System Enhancement Program (ALS-SEP)

FOREIGN MILITARY SALES
None

CONTRACTORS
Wolf Coach, Inc., an L-3 Communications company (Auburn, MA)

Armed Reconnaissance Helicopter (ARH)

Provides a robust reconnaissance and security capability for the joint combined arms air-ground maneuver team, improving the commander's ability to concentrate superior combat power against the enemy.

INVESTMENT COMPONENT

- **Modernization**
- Recapitalization
- Maintenance

Description & Specifications

The Armed Reconnaissance Helicopter (ARH) provides an easily deployable, rapidly reconfigurable, reconnaissance and security capability that can be employed immediately upon arrival in theater. The ARH is a dual-crew station, single-pilot-operable aircraft leveraging the Bell Helicopter 407 commercial platform.

The ARH's standard armed reconnaissance configuration includes the sensor assembly, active and passive countermeasures, external weapon systems, and communication suite. The ARH will communicate on the battlefield with Army, joint, and coalition forces. The ARH will engage short- and long-range mobile/non-mobile targets with a combination of .50 caliber, 2.75" rockets and/or Hellfire missiles. The armed reconnaissance capability allows the maneuver commander to "see first/shoot first" with a manned system that is well forward, scouting the location and disposition of enemy forces and relevant terrain.

Program Status

- **4QFY05:** Milestone B acquisition decision
- **4QFY05:** Contract awarded to Bell Helicopter
- **4QFY05:** System requirements review
- **1QFY06:** Systems functional review
- **1QFY06:** Preliminary design review
- **4QFY06:** First flight

Recent and Projected Activities

- **2QFY07:** Critical design review
- **2QFY07:** Limited user test
- **3QFY07:** Milestone C acquisition decision
- **4QFY08:** Initial operational test and evaluation

ACQUISITION PHASE

Concept & Technology Development | **System Development & Demonstration** | Production & Deployment | Operations & Support

Armed Reconnaissance Helicopter (ARH)

FOREIGN MILITARY SALES
None

CONTRACTORS
Bell Helicopter Textron (Ft. Worth, TX)
FLIR Systems, Inc. (Portland, OR)
Rockwell Collins (Cedar Rapids, IA)
Honeywell ES&S (Phoenix, AZ)

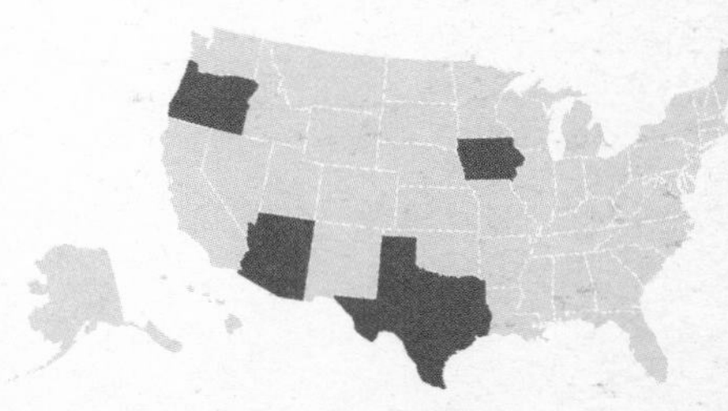

Armored Knight

Performs 24-hour terrain surveillance, target acquisition, target location, and fire support Combat Observation Lasing Team mission execution in Heavy and Infantry Brigade Combat Teams.

INVESTMENT COMPONENT

- **Modernization**
- Recapitalization
- Maintenance

Description & Specifications

The M1200 Armored Knight provides precision strike capability by locating and designating targets for both ground- and air-delivered laser-guided ordnance and conventional munitions. It replaces the M707 Knight High Mobility Multipurpose Wheeled Vehicle (HMMWV base) and M981 Fire Support Team Vehicles used by Combat Observation Lasing Teams (COLTs) in both the Heavy and Infantry Brigade Combat Teams. It operates as an integral part of the brigade reconnaissance element, providing COLT and fire support mission planning and execution.

The Armored Knight is a M1117 Armored Security Vehicle (ASV)-based platform providing enhanced survivability and maneuverability. The system includes a full 360-degree armored cupola and integrated Knight Mission Equipment Package (MEP) that is common with the M7 BFIST/M707 Knight and the Stryker Fire Support Vehicle. The common components are:

- FS3 mounted sensor
- Targeting Station Control Panel
- Mission Processor Unit
- Inertial Navigation Unit
- Defense Advanced Global Positioning System Receiver (DAGR)
- Power Distribution Unit
- Stand-alone Computer Unit

Other Armored Knight specifications include the following:

- Crew: Three COLT members
- Combat loaded weight: Approximately 15 tons
- Maximum speed: 63 mph
- Cruising range: 440 miles
- Target location accuracy: <20 meters CEP

Program Status

- **4QFY06:** Approval to begin production of Armored Knight to replace the HMMWV chassis Knight

Recent and Projected Activities

- **FY07:** Complete design/integration/verification and begin production
- **4QFY07:** Provide five Armored Knights in support of 10th Mountain Division ONS
- **3Q–4QFY08:** Delivery and fielding to 1st Cavalry, 3rd Infantry Division, 1st Infantry Division, and 82nd Airborne Division
- **FY09:** Continue delivery and fielding

ACQUISITION PHASE

Concept & Technology Development | **System Development & Demonstration** | **Production & Deployment** | Operations & Support

Armored Knight

Crew: 3
GVW: 29,560 lbs.
Speed: 63 mph
Range: 440 miles

FOREIGN MILITARY SALES
None

CONTRACTORS

Prime:
Precision targeting systems production/vehicle integration:
DRS Sustainment Systems, Inc. (DRS-SSI) (St. Louis, MO; West Plains, MO)

Subcontractors:
System cables:
DRS Laurel Technologies (Johnstown, PA)
Common display unit:
DRS Tactical Systems (Melbourne, FL)
Slip ring:
Airflyte Electronics Co. (Bayonne, NJ)
Targeting station control panel:
Oppenheimer (Horsham, PA)
Major GFE/GFM contractors:
M1117 ASV Hull:
Textron Marine & Land Systems (New Orleans, LA)

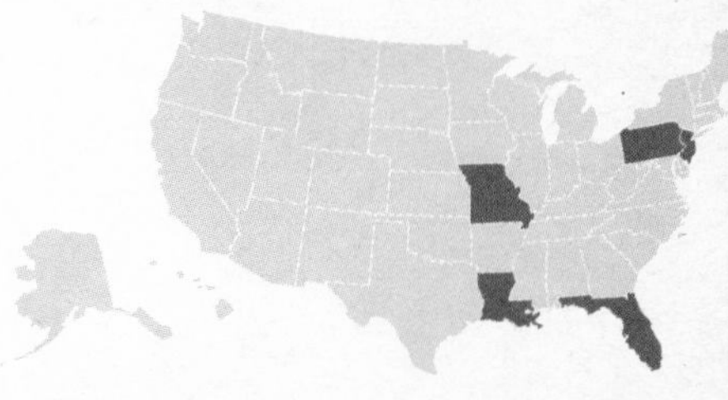

Armored Security Vehicle (ASV)

Supports the entire spectrum of military police missions and protects convoys in hostile areas.

INVESTMENT COMPONENT

Modernization
Recapitalization
Maintenance

Description & Specifications

The M1117 Armored Security Vehicle (ASV) is a turreted, lightly armored, all-wheel drive vehicle that supports military police missions, such as rear area security, law and order operations, convoy protection, battlefield circulation, and enemy prisoner of war operations, over the entire spectrum of war and operations other than war.

The ASV provides protection to the crew compartment, gunner's station, and the ammunition storage area. The turret is fully enclosed with both an MK-19 40mm grenade machine gun and a .50-caliber machine gun, and a multi-salvo smoke grenade launcher. The ASV provides ballistic, blast, and overhead protection for its four-person crew. The ASV has a payload of 3,360 pounds and supports Army transformation with its 400 miles plus range, top speed of nearly 70 miles per hour, and C-130 deployability.

The ASV is the Army's chosen convoy protection platform, and production is being accelerated to meet this mission as well as its military police requirement mission.

Program Status

- **FY05–FY11:** Continued fielding to support military police companies and convoy protection

Recent and Projected Activities

- **FY07–FY09:** Produce and field 1,100 vehicles

ACQUISITION PHASE

Concept & Technology Development | System Development & Demonstration | Production & Deployment | Operations & Support

Armored Security Vehicle (ASV)

FOREIGN MILITARY SALES
ASV variant delivered to Iraq

CONTRACTORS
Textron Marine & Land Systems
(New Orleans, LA)

Army Airborne Command and Control System (A2C2S)

Enables commanders from the brigade through theater level to perform battle command functions from an airborne platform through continuous situational awareness and robust communications.

INVESTMENT COMPONENT

- **Modernization**
- Recapitalization
- Maintenance

DESCRIPTION AND SPECIFICATIONS

The Army's UH-60L Black Hawk (and newer models) hosts the Army Airborne Command and Control System (A2C2S). The A2C2S consists of all aircraft modifications, antennas, and communications and electronic components enabling battle command on an airborne platform. The A2C2S provides operator workstations, computer and network systems, and necessary security devices to host and support the digital battle command process. The A2C2S is capable of operations on both SIPR and NIPR networks and hosts the Army Battle Command System programs for situational awareness, maneuver, effects, and intelligence data and collaboration. Data transport is provided by the Ku-band Satellite terminal and Blue Force Tracking-Aviation. A2C2S voice communications consist of Combat Net Radios and High Frequency SATCOM with DAMA and IRIDIUM. The One System Remote Video Terminal provides real-time video and telemetry from the Shadow Tactical Unmanned Aerial Vehicle System. Civil Land Mobile Radio enhances communications interoperability with state and local agencies in a homeland defense or disaster situation.

The A2C2S is fielded to the Command Aviation Companies (CAC) of the General Support Aviation Battalions in the Combat Aviation Brigades and Theater Aviation Brigades of the Modular Force.

Program Status

- **FY05:** A2C2S v1.0 fielded to 4th Infantry Division (ID), 101st Air Assault (AA), and 10th Mountain Division
- **FY06:** A2C2S v1.0 fielded to 25th ID and 1st Cavalry Division, and 36th Combat Aviation Brigade (CAB)

Recent and Projected Activities

- **FY07:** A2C2S v1.1 fielded to 82nd Airborne, 1st ID, 3rd ID, 101st AA, and 12th CAB
- **FY07:** Complete developmental and operational test activities
- **1QFY08:** Full rate production decision; fielding to the Active Duty and National Guard CABs
- **FY09:** Fielding to National Guard CABs and to Theater Aviation Brigades

ACQUISITION PHASE

- Concept & Technology Development
- System Development & Demonstration
- **Production & Deployment**
- Operations & Support

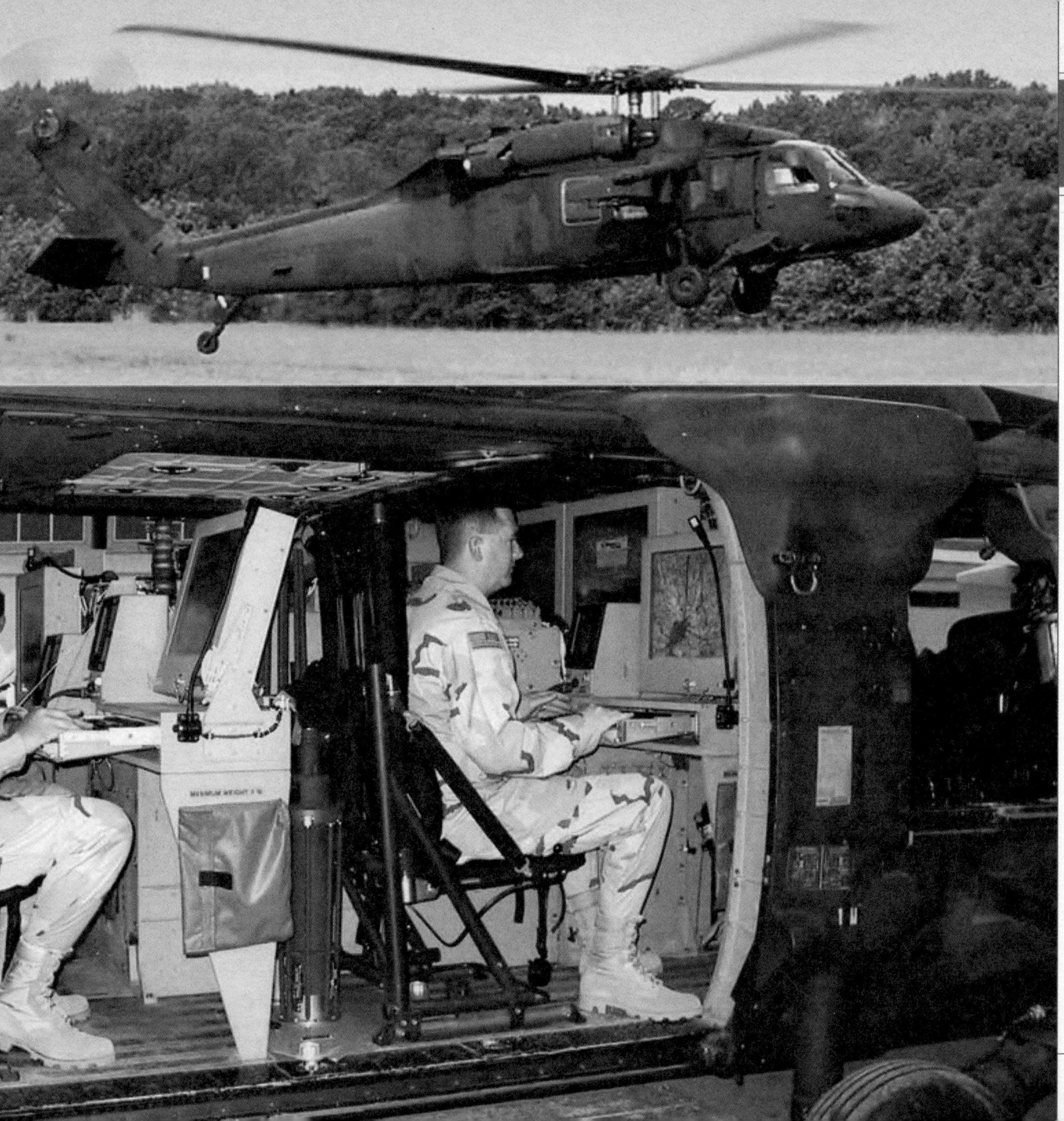

Army Airborne Command and Control System (A2C2S)

FOREIGN MILITARY SALES
None

CONTRACTORS
Joint Venture Yulista SES-I (Huntsville, AL)
Boeing (Philadelphia, PA)
Program support:
Prototype Integration Facility (Redstone Arsenal, AL)
Aviation Applied Technologies Directorate (Ft. Eustis, VA)

Army Data Distribution System (ADDS)/Enhanced Position Location Reporting System (EPLRS)

Provides embedded situational awareness and position navigation for the warfighter using mobile, digitized, wireless data communications.

INVESTMENT COMPONENT

- **Modernization**
- Recapitalization
- Maintenance

Description & Specifications

The Enhanced Position Location Reporting System (EPLRS) supports the Army's digitized divisions and Stryker Brigade Combat Teams (SBCTs). EPLRS is the backbone of the Army's Tactical Internet, providing data distribution and position navigation services in near real-time for the warfighter at brigade-and-below level, in support of Army Battle Command Systems (ABCS) and the Force XXI Battle Command Brigade-and-Below (FBCB2) program.

EPLRS consists of a network control station and the EPLRS radio, which can be configured as a manpack unit, a surface vehicle unit, and an airborne vehicle unit. EPLRS uses a time-division, multiple-access communications architecture to avoid transmission contention. In addition, it uses frequency hopping and error detection and correction with interleaving. Spread-spectrum technology provides jamming resistance.

EPLRS is designed to be used as a common system for the Army, Air Force, Navy, and Marine Corps. Improvements to EPLRS include message reliability, more efficient available bandwidth, and field-programmable software.

Program Status

- **3QFY04:** Contract award for 711 EPLRS radios
- **FY06:** Fielding ongoing to SBCT 5

Recent and Projected Activities

- **FY05–07:** Continue fielding to Air National Guard Air Defense Artillery Units, SBCTs 4-6, 3rd Armored Cavalry Regiment, and III Corps Troops
- **FY05–07:** Field EPLRS retrofit kits (increases throughput to 288 Kbps) to 1st Cavalry Division and 3rd and 4th Infantry Divisions

CQUISITION PHASE

- oncept & Technology Development
- System Development & Demonstration
- **Production & Deployment**
- Operations & Support

Army Data Distribution System (ADDS)/Enhanced Position Location Reporting System (EPLRS)

FOREIGN MILITARY SALES
None

CONTRACTORS
Radio design/production:
Raytheon (Fullerton, CA; Forest, MS; Ft. Wayne, IN; Garland, TX)
Engineering support:
British Aerospace Engineering (BAE) Systems (West Long Branch, NJ)
Fielding:
Innolog (Wall Township, NJ)
Engineering Professional Services (EPS) (Shrewsbury, NJ)

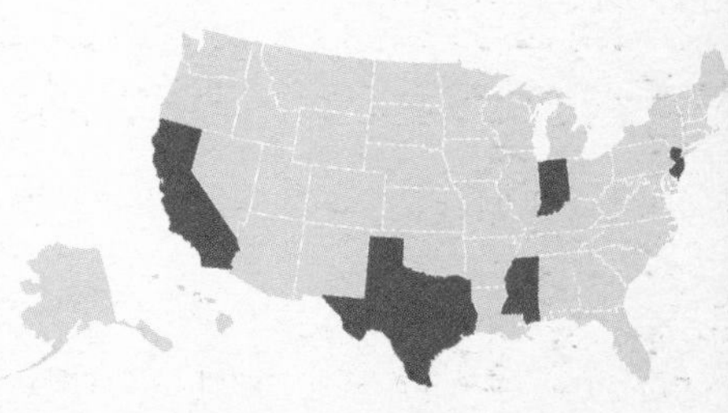

Army Key Management System (AKMS)

Provides planners and operators with automated, secure communications at both the theater/tactical and strategic/sustaining base levels.

INVESTMENT COMPONENT

- **Modernization**
- Recapitalization
- Maintenance

Description & Specifications

The Army Key Management System (AKMS) automates the functions of communications security (COMSEC) key management control and distribution, electronic counter-countermeasures generation and distribution, and signal operating instructions management. AKMS supports joint interoperability, while limiting adversarial access to and decreasing the vulnerability of Army command, control, communications, computers, and intelligence systems.

AKMS is composed of the following three elements:

1. Local COMSEC management software (LCMS), which performs COMSEC accounting and electronic key generation and distribution.

2. Automated Communications Engineering Software (ACES), which is the frequency management portion of AKMS and will be designated by the Military Communications Electronics Board as the joint standard for use by all services in development of frequency management and cryptographic net planning. ACES will replace the legacy Revised Battlefield Electronic Communications-Electronic Operating Instructions System (RBECS) and will become the joint electronic interface to all spectrum management, Integrated System Control, Spectrum XXI, RBECS, Operational Tasking Command, Air Tasking Order (ATO), and Space ATO workstations.

3. The Simple Key Loader, which replaces the Data Transfer Device (DTD). The small design allows easy key transfers and provides the interface among LCMS, ACES, and End Crypto units.

Program Status

- **2QFY06:** Procure and field Simple Key Loaders
- **3QFY06:** Completion of ACES fieldings

Recent and Projected Activities

- **2QFY07:** Procure and field Simple Key Loaders; future ACES Block II and DTD/Simple Key Loader upgrades
- **FY07–09:** Future ACES Block II and DTD/Simple Key Loader upgrades

ACQUISITION PHASE

Concept & Technology Development | System Development & Demonstration | **Production & Deployment** | Operations & Support

Army Key Management System (AKMS)

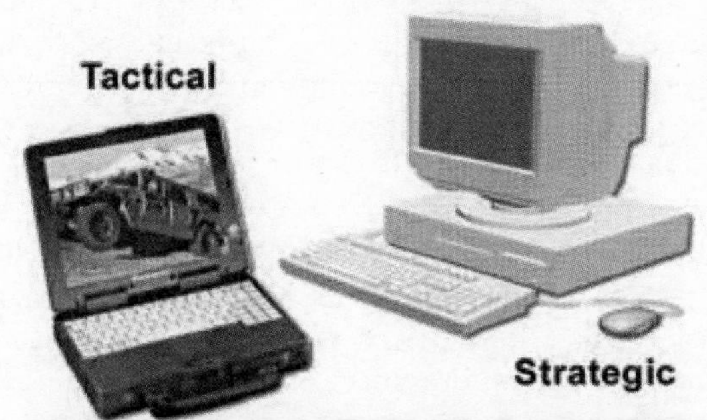

FOREIGN MILITARY SALES
None

CONTRACTORS
Software:
Sierra Nevada Corp. (Sparks, NV)
SAIC (San Diego, CA)
Information Systems Support, Inc. (Tinton Falls, NJ)
Inter4 (San Francisco, CA)
Sypris (Tampa, FL)
CSS (Augusta, GA)

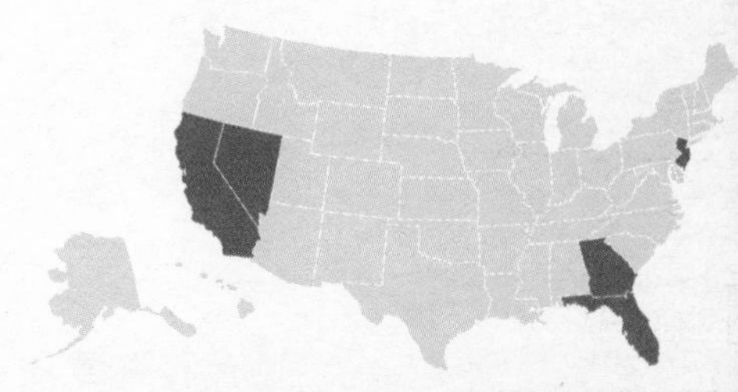

Army Tactical Missile System (ATACMS)

Provides the joint forces commander with long-range, all-weather, around-the-clock fires to attack area and point targets with precision, while minimizing collateral damage.

INVESTMENT COMPONENT

- Modernization
- Recapitalization
- Maintenance

Description & Specifications

The Army Tactical Missile System (ATACMS) is a ground-launched missile system that consists of surface-to-surface guided missiles with variants containing anti-personnel/ anti-materiel and unitary warheads. ATACMS missiles are fired from the M270A1 Multiple Launch Rocket System (MLRS) and the High Mobility Artillery Rocket System (HIMARS) launchers. ATACMS variants can engage targets with Global Positioning System accuracy at ranges well beyond the capability of existing cannons and rockets, and in situations when air-delivered munitions are not available. Operations include Joint Suppression of Enemy Air Defense (JSEAD) and shaping and shielding fires. Targets include air defense systems, logistics elements, and command, control, and communications complexes.

Block I and Block IA missiles have anti-personnel/anti-materiel warheads. Block I has a range of 165 kilometers, and Block IA has a range of 300 kilometers.

Quick Reaction Unitary (QRU) missiles have a 500-pound, high-explosive warhead with a point-detonating fuze and a range of 270 kilometers. These missiles were produced to meet an urgent operational requirement for the Global War on Terror.

ATACMS Unitary Product Improvement Program missiles are in development with a tri-mode fuze (point-detonating, delay, and aerial burst) and an improved 500-pound high-explosive warhead. The Product Improvement Program missiles will begin production in FY09.

Program Status

- **1Q–4QFY06:** Delivery of QRU missiles
- **2QFY06:** QRU production contract award for 48 missiles
- **3QFY06:** ATACMS Unitary Product Improvement Program research, development, test, and evaluation contract award
- **4QFY06:** QRU supplemental contract award for 50 missiles

Recent and Projected Activities

- **1Q–4QFY07:** Delivery of QRU missiles
- **2QFY07:** QRU production contract award
- **1Q–4QFY08:** Delivery of QRU missiles
- **2QFY08:** QRU production contract award
- **1Q–4QFY09:** Delivery of QRU missiles
- **2QFY09:** ATACMS Unitary production contract award

ACQUISITION PHASE

Concept & Technology Development | System Development & Demonstration | Production & Deployment | Operations & Support

Army Tactical Missile System (ATACMS)

FOREIGN MILITARY SALES
Bahrain, Greece, South Korea, Turkey

CONTRACTORS
Prime:
Lockheed Martin (Dallas, TX; Horizon City, TX)
Guidance section:
Honeywell (Clearwater, FL)
Fuze:
Kaman (Middleton, CT)
Boat tail:
Goodrich (Vergennes, VT)
Nose cone:
Spincraft (New Berlin, WI)
Rocket motor and case:
Aerojet (Camden, AR)

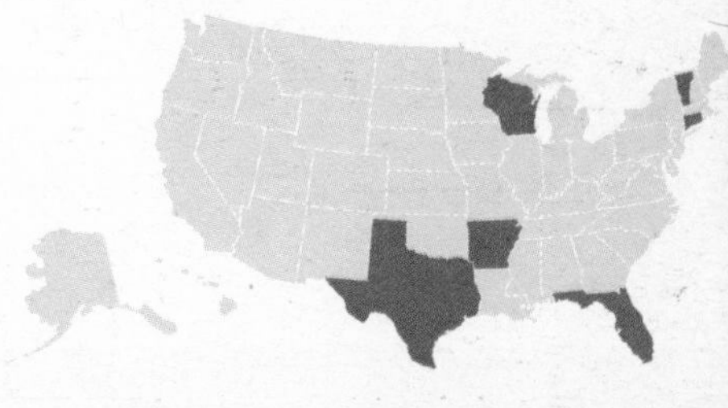

Artillery Ammunition

Provides field artillery forces with modernized, highly precise munitions to destroy, neutralize, or suppress the enemy by cannon fire.

INVESTMENT COMPONENT

Modernization

Recapitalization

Maintenance

Description & Specifications

The Army's artillery ammunition program includes 75mm (used for ceremonies and simulated firing), 105mm, and 155mm projectiles and their associated fuzes and propelling charges.

Semi-fixed ammunition for short and intermediate ranges, used in 105mm howitzers, is characterized by adjusting the number of multiple propelling charges. Semi-fixed ammunition for long ranges contains a single bag of propellant optimized for obtaining high velocity and is not adjustable. The primer is an integral part of the cartridge case, and is located in the base. All 105mm cartridges are issued in a fuzed or unfuzed configuration. Both cartridge configurations are packaged with propellant.

Separate-loading ammunition, used in 155mm howitzers, has separately issued projectiles, fuzes, propellants, and primers, which are loaded into the cannon separately.

The artillery ammunition program includes fuzes for cargo-carrying projectiles, such as smoke, illumination, dual purpose improved conventional munitions, and bursting projectiles, such as high explosive. This program also includes bag propellant for the 105mm semi-fixed cartridges and modular artillery charge system (MACS) for 155mm howitzers.

The Precision Guidance Kit is a low-cost system that will improve the accuracy of conventional 105mm and 155mm artillery projectiles.

Program Status

- **3QFY05:** MACS M232A1 high zone propelling charge optimized for the 155mm/39-caliber howitzer and type classified

Recent and Projected Activities

- **FY07:** Precision Guidance Kit begins system development and demonstration

ACQUISITION PHASE

Concept & Technology Development | System Development & Demonstration | Production & Deployment | Operations & Support

Artillery Ammunition

FOREIGN MILITARY SALES
Multiple foreign purchases

CONTRACTORS
General Dynamics Ordnance and Tactical Systems-Scranton Operations (Scranton, PA)
SNC Technologies (LeGardeur, Canada)
American Ordnance (Middletown, IA)
Alliant Techsystems (Janesville, WI)
Armtec Defense (Palm Springs, CA)

Aviation Combined Arms Tactical Trainer (AVCATT)

Enables Army aviation units to rehearse and participate in a unit-collective and combined-arms simulated battlefield environment through networked simulation training.

INVESTMENT COMPONENT

Modernization

Recapitalization

Maintenance

Description & Specifications

The Aviation Combined Arms Tactical Trainer (AVCATT) is a reconfigurable, transportable, combined arms virtual training simulator that provides current and Future Force aviation commanders and units a dynamic, synthetic instructional environment. It meets institutional, organizational, and sustainment aviation training requirements for Active and Reserve Army aviation units worldwide and enables geographic-specific mission rehearsals before real-world mission execution.

AVCATT is a critical element of the Combined Arms Training Strategy. It is distributive interactive simulation- and high level architecture-compliant, and is compatible and interoperable with other synthetic environment systems. AVCATT supports role-player and semi-automated blue and opposing forces.

The AVCATT single suite of equipment consists of two mobile trailers housing six reconfigurable networked simulators that support the Apache, Apache Longbow, Kiowa Warrior, Chinook, Black Hawk, and the Armed Reconnaissance Helicopter (ARH) (FY09). An after-action review theater and battle master control station is also provided as part of each suite.

AVCATT enables realistic, high-intensity collective and combined arms training to aviation leadership, staff members and units, improving overall aviation task force readiness. AVCATT builds and sustains training proficiency on mission-essential tasks through crew and individual training by supporting aviation collective tasks, including the following:

- Armed reconnaissance (area, zone, route)
- Deliberate attack
- Covering force operations
- Downed aircrew recovery operations
- Joint air attack team
- Hasty attack
- Air assault operations

AVCATT is fully mobile, capable of using commercial and generator power, and is transportable worldwide.

Program Status

- **3QFY06:** Fielding of 10 suites completed; includes support to Army National Guard as well as U.S. forces in Germany, Korea, and Hawaii
- **1QFY07:** Production contract for suites 16–20

Recent and Projected Activities

- **2QFY07:** Initiate development of ARH baseline
- **3QFY07:** Initiate development of Common Avionics Architectural Systems (CAAS) for future concurrency airframes
- **4QFY07:** Field the first Longbow Block II concurrency upgrade
- **1QFY08:** Production contract for suites 21–23; field the first classified operational suite
- **1QFY09:** Planned hardware and software integration of new image generator upgrade and helmet mounted display

ACQUISITION PHASE

Concept & Technology Development | System Development & Demonstration | Production & Deployment | Operations & Support

Aviation Combined Arms Tactical Trainer (AVCATT)

FOREIGN MILITARY SALES
None

CONTRACTORS
L-3 Communications (Arlington, TX)

AVCATT-A
2 Trailer Suite

AVCATT-AAR
AVCATT-BMC
(left to right)

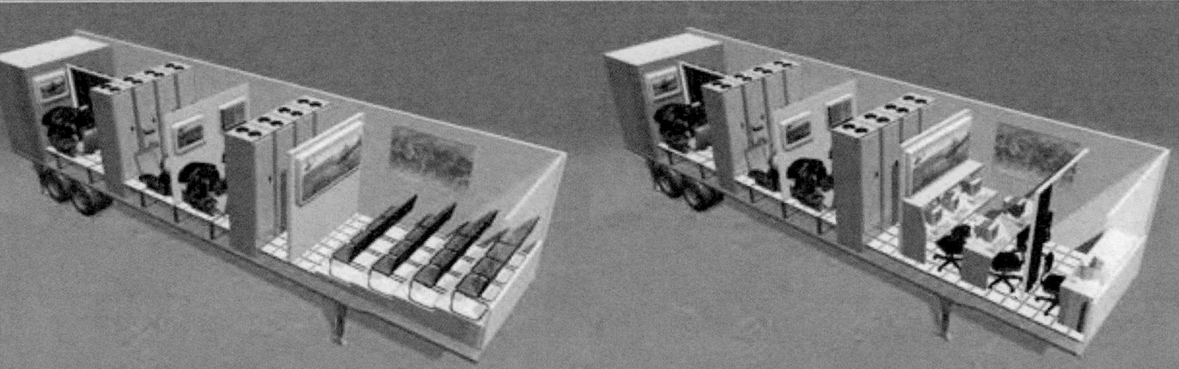

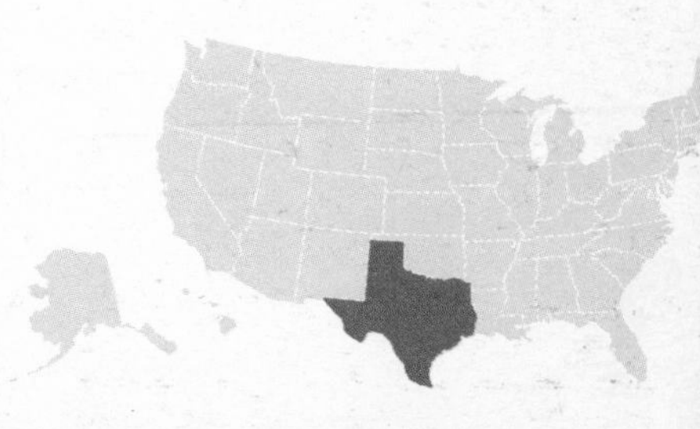

Battle Command Sustainment Support System (BCS3)

Provides current, map-based information in a user-friendly operating environment for logistics decision support systems for warfighters on the battlefield.

INVESTMENT COMPONENT

- **Modernization**
- Recapitalization
- Maintenance

Description & Specifications

The Battle Command Sustainment Support System (BCS3) is the Army's logistics command and control system solution for U.S. land forces. It provides commanders at all echelons the capability to execute end-to-end distribution and deployment management, and, as such, has been adopted and integrated into joint and strategic logistics command and control processes.

By connecting logisticians, BCS3 provides an effective and efficient means to gather and integrate asset and in-transit information to manage distribution and deployment missions. Logisticians can visualize movements of personnel, supplies, and transportation assets. This information enables warfighters to target, access, scale, and tailor critical logistics information in near-real time.

BCS3 leverages commercial software technology, merging previous and current system functionality. Using a 6-pound, commercial laptop, logisticians can visualize items throughout the supply chain. With its map-based display, BCS3 allows users to project supplies for different scenarios and moves data seamlessly from unclassified to classified networks in a National Security Agency-certified manner.

Commanders can see the logistical situation and plan, rehearse, train, and execute on a single system. Its network-centric architecture meets the full spectrum of operations at echelons brigade and above. BCS3 interfaces with the Army Battle Command System (ABCS) battlefield functional areas, as well as external higher and lower command and communications systems.

Program Status

- **3QFY06:** Successful completion of ABCS and Enabler Systems Test
- **3QFY06:** Fielding to III Corps and 13th Sustainment Command
- **4QFY06:** Fielding of BCS3 version 11.23.5E
- **4QFY06:** Fielding to 3rd Infantry Division
- **1QFY07:** Implementation of regional support for BCS3
- **1QFY07:** Fielding to 173rd Brigade Combat Team and 12th Combat Aviation Brigade
- **1QFY07:** System integration, testing, and fielding of updates

Recent and Projected Activities

- **2QFY07:** Fielding to 1st Sustainment Command
- **2QFY07:** ABCS system integration
- **3QFY07:** Fielding to 101st Airborne Division
- **4QFY07:** Fielding to 2nd Infantry Division
- **1QFY08:** Fielding to 1st Armored Division
- **2QFY08:** Fielding to 1st Infantry Division

AQUISITION PHASE

- Concept & Technology Development
- System Development & Demonstration
- **Production & Deployment**
- Operations & Support

Battle Command Sustainment Support System (BCS3)

FOREIGN MILITARY SALES
None

CONTRACTORS
Tapestry Solutions (San Diego, CA)
Northrop Grumman (Carson, CA)
L-3 Communications (Chantilly, VA)
Lockheed Martin (Tinton Falls, NJ)
Wexford Group International (Vienna, VA)

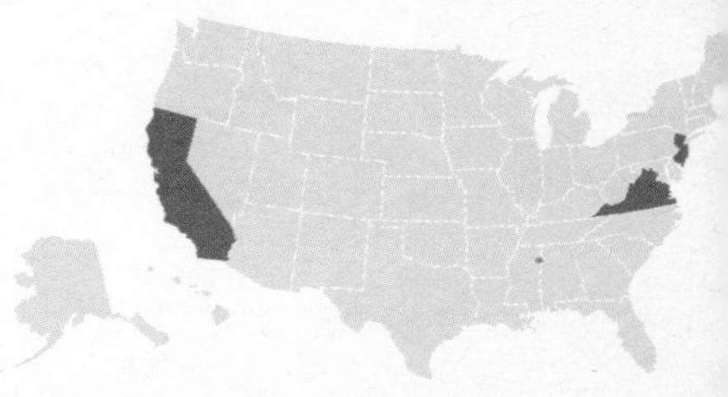

Black Hawk/UH-60

Provides air assault, general support, aero-medical evacuation, command and control, and special operations support.

INVESTMENT COMPONENT

Modernization

Recapitalization

Maintenance

Description & Specifications

The Black Hawk/UH-60 is the Army's utility tactical transport helicopter. Because of dramatic improvements in troop capacity and cargo lift capability, the versatile Black Hawk has enhanced the overall mobility of the Army. It will serve as the Army's utility helicopter in the Future Force.

There are four basic versions of the UH-60: the original UH-60A; the UH-60L, which has greater gross weight capability, higher cruise speed, rate of climb and external load; the UH-60M, which includes the improved GE-701D engine, providing higher cruise speed, rate of climb, and internal load than the UH-60A and L versions; and the UH-60M Pre-Planned Product Improvements (P3I) Upgrade, which includes Composite Tail Cone and Drive Shafts, Common Avionics Architecture System, Fly-By-Wire, and Full Authority Digital Engine Control when used with the GE-701D Engine. The UH-60M program will incorporate a digitized cockpit and improved handling characteristics, and will extend the service life of the system.

On the asymmetric battlefield, the Black Hawk provides the commander the agility to get to the fight more quickly and to mass effects throughout the battlespace across the full spectrum of conflict. A single Black Hawk can transport an entire 11-person, fully equipped infantry squad faster than predecessor systems and in most weather conditions. The Black Hawk can reposition a 105mm howitzer, its crew of six, and up to 30 rounds of 105mm ammunition in a single lift. The aircraft's critical components and systems are armored or redundant, and its airframe is designed to protect crew and passengers by crushing progressively on impact.

Program Status

- **1QFY07:** UH-60 operational testing
- **1QFY07:** Critical design review of UH-60M P3I upgrade

Recent and Projected Activities

- **3QFY07:** UH-60M full rate production decision
- **3QFY07:** Award multi-year production contract for UH-60M
- **1QFY08:** First flight of UH-60M P3I upgrade

ACQUISITION PHASE

Concept & Technology Development | System Development & Demonstration | Production & Deployment | Operations & Support

Black Hawk/UH-60

FOREIGN MILITARY SALES
None

CONTRACTORS
United Technologies (Stratford, CT)
General Electric (Lynn, MA)
Rockwell Collins (Cedar Rapids IA)
ATI Firth Sterling (Madison, AL)
Goodrich-Hella (Oldsmar, FL)

	UH-60A	UH-60L	UH-60M	UH-60M P31 Upgrade
MAX GROSS WEIGHT (pounds):	20,250	22,000	22,000	22,000
CRUISE SPEED (knots):	149	150	152	152
RATE CLIMB (feet per minute):	814	1,315	1,646	1,646
ENGINES (2 each):	GE-700	GE-701C	GE-701D	GE-701D
EXTERNAL LOAD (pounds):	8000	9,000	9,000	9,000
INTERNAL LOAD (troops/pounds):	11/2, 640	11/2, 640	11/3, 190	11/3, 190
CREW:	two pilots, two crew chiefs			
ARMAMENT:	two 7.62mm machine guns			

Bradley Upgrade

Provides infantry and cavalry fighting vehicles with digital command and control capabilities, significantly increased situational awareness, enhanced lethality and survivability, and improved sustainability and supportability.

INVESTMENT COMPONENT

- Modernization
- Recapitalization
- Maintenance

Description & Specifications

The Bradley M2A3 Infantry/M3A3 Cavalry Fighting Vehicle (IFV/CFV) features two second generation, forward looking infrared (FLIR) sensors—one in the Improved Bradley Acquisition Subsystem (IBAS), the other in the Commander's Independent Viewer (CIV). These systems provide "hunter-killer target handoff" capability with ballistic fire control.

The Bradley A3 also has embedded diagnostics and an integrated combat command and control (IC3) digital communications suite hosting a Force XXI Battle Command Brigade-and-Below (FBCB2) package with digital maps, messages, and friend/foe situational awareness. The Bradley's position navigation with GPS, inertial navigation, and enhanced squad situational awareness includes a squad leader display integrated into vehicle digital images and IC3.

Program Status

- **Current:** Fielded to the 3rd Infantry Division (ID) and the 3rd Armored Cavalry Regiment (ACR); 3rd ID is receiving ODS Bradleys and 3rd ACR is receiving A3 Bradleys
- **Current:** Bradley conversions continue for both the active Army and the ARNG to meet the Army's modularity goals; A3 Bradleys in full-rate production for several years
- **FY04:** A3 production, started in 2QFY06, incorporated CMED, and an upgrade to the two FLIR systems in Block I, reducing quantity of LRU components, and support and acquisition costs
- **3QFY06:** Bradley ODS fielded to the 3rd ID
- **4QFY06:** Bradley ODS fielded to the 3rd ID; FY06 Reset contract awarded for 72 A3 and 23 ODS
- **1QFY07:** Bradley ODS fielded to the 3rd ID; Bradley A3 fielded to 3rd ACR; Reset contract awarded for 425 A3 and 118 ODS

Recent and Projected Activities

- **2QFY07:** Bradley ODS fielded to the 116th ID ARNG; Bradley A3 fielded to 3rd ACR
- **3QFY07:** Bradley A3 fielded to the EBCT, 1st Armored Division with Spin-Out 1 technologies; Bradley ODS fielded to the 81st WA ARNG
- **4QFY07:** Bradley A3 fielded to 4th ID; Bradley ODS fielded to the 278th TN ARNG
- **1QFY08:** Bradley A3 fielded to 1st Armored Division; Bradley ODS fielded to the ARNG
- **2QFY08:** Bradley A3 fielded to 1st Cavalry Division; Bradley ODS fielded to the ARNG
- **3QFY08:** Bradley A3 fielded to 1st Cavalry Division
- **4QFY08:** Bradley A3 fielded to 1st Armored Division; Bradley ODS fielded to the 1st ID; Bradley ODS fielded to the ARNG
- **1QFY09:** Bradley A3 fielded to 1st Armored Division; Bradley ODS fielded to the ARNG

ACQUISITION PHASE

- Concept & Technology Development
- System Development & Demonstration
- Production & Deployment
- Operations & Support

Bradley Upgrade

FOREIGN MILITARY SALES
None

CONTRACTORS
British Aerospace Engineering (BAE) Systems (San Jose, CA)
Raytheon (McKinney, TX)
DRS Technologies (Melbourne, FL)
EFW (Ft. Worth, TX)
L-3 Communications (Muskegon, MI)

LENGTH: 21.5 feet
WIDTH: 10.75 feet without armor tiles; 11.83 feet with armor tiles
HEIGHT: 11.8 feet
WEIGHT: 67,000 pounds combat loaded; 72,000 pounds with armor tiles
POWER TRAIN: 600 hp Cummins VTA-903T diesel engine with GM-Allison HMPT-500-3EC hydro-mechanical automatic transmission
CRUISING RANGE: 250 miles
ROAD SPEED: 38 mph
CREW: M2A3: 10 (3 crew; 7 dismounts); M3A3: 5 (3 crew; 2 dismounts)
VEHICLE ARMAMENT: 25mm Bushmaster cannon; TOW II missile system; 7.62mm M240C machine gun

CURRENT MODELS/VARIANTS:
- M2/M3 A2
- M2/M3 A2 Operation Desert Storm (ODS)
- M2/M3 A3
- M2/M3 A2 ODS-E (Engineer Vehicle)
- Bradley Commander's Vehicle (BCV)
- M7 ODS Bradley Fire Support Team (BFIST)
- M3A3 Bradley Fire Support Team (BFIST)

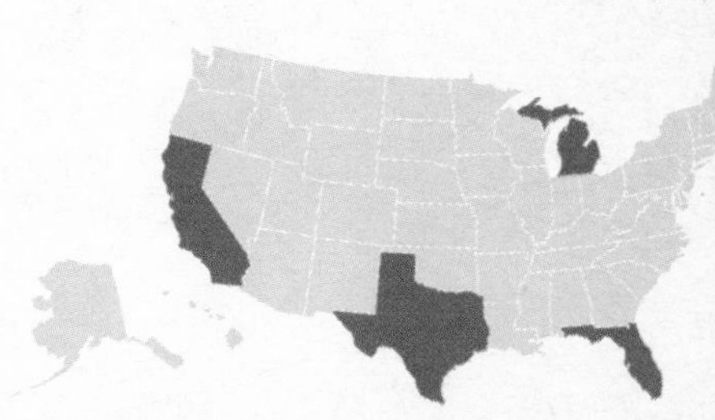

Chemical Biological Protective Shelter (CBPS)

Enables medical personnel to treat casualties without the encumbrance of individual protective clothing and equipment in a highly mobile, easy-to-use, and self-contained, chemical-biological hardened facility.

INVESTMENT COMPONENT

- Modernization
- Recapitalization
- Maintenance

Description & Specifications

The Chemical Biological Protective Shelter (CBPS) is a highly mobile, self-contained system designed to replace the M51 Collective Protection Shelter. CBPS consists of a Lightweight Multi-purpose Shelter (LMS) mounted on an expanded capacity variant (ECV) High Mobility Multi-Purpose Wheeled Vehicle (HMMWV) and a 300-square-foot, airbeam-supported soft shelter. CBPS provides a contamination-free, environmentally controlled working area for medical, combat service, and combat service support personnel to obtain relief from the need to wear chemical-biological protective clothing for 72 hours of operation.

All ancillary equipment required to provide protection, except the generator, is mounted within the shelter. Medical equipment and crew gear are transported inside of the LMS or by a towed high-mobility trailer.

CBPS will be assigned to the trauma treatment teams/squads of the maneuver battalions, the medical companies of the forward and division support battalions, non-divisional medical treatment teams/squads, division and corps medical companies, and the forward surgical teams.

Program Status

- **2QFY06:** Indefinite Delivery/ Indefinite Quantity contract awarded for electrically powered CBPS variants

Recent and Projected Activities

- **FY07–09:** Procure and field systems

ACQUISITION PHASE

Concept & Technology Development | System Development & Demonstration | Production & Deployment | Operations & Support

Chemical Biological Protective Shelter (CBPS)

FOREIGN MILITARY SALES
None

CONTRACTORS
DRS Technologies (Parsippany, NJ)
Smiths Detection, Inc. (Edgewood, MD)

Chemical Demilitarization

Safely destroys all chemical warfare and related materiel while ensuring maximum protection for the public, workers, and the environment.

INVESTMENT COMPONENT

- Modernization
- Recapitalization
- Maintenance

Description & Specifications

The Chemical Materials Agency (CMA) is responsible for the destruction of chemical agents and munitions, disposal facility design, construction, systemization, operations, and closure, except at the stockpile locations in Colorado and Kentucky. CMA is also responsible for emergency preparedness activities at chemical weapons storage depots; disposal of binary chemical munitions and non-stockpile chemical materiel; destruction of former chemical weapons production facilities; assessment and destruction of recovered chemical warfare materiel; and provision of all DoD-approved support to international chemical demilitarization programs.

Program Status

- **3QFY06:** Destroyed last of U.S. inventory of DF binary precursor to GB nerve agent at Pine Bluff, AR.
- **4QFY06:** Began processing VX nerve agent-filled munitions at Anniston, AL
- **4QFY06:** Began processing mustard-filled munitions at Tooele, UT
- **4QFY06:** Completed processing GB-filled bombs and rockets at Umatilla, OR
- **4QFY06:** Destroyed over 40 percent of the original U.S. chemical stockpile, including more than 1.7 million munitions
- **4QFY06:** Destroyed last of U.S. inventory of QL binary precursor to VX at Pine Bluff, AR
- **1QFY07:** Continue closure activities at Aberdeen, MD
- **1QFY07:** Complete destruction of recovered Chemical Agent Identification Sets at Pine Bluff, AR

Recent and Projected Activities

- **2QFY07:** Begin destruction of binary neutralent in the Wet Air Oxidation system at Texas Molecular in Deer Park, TX
- **3QFY07:** Achieve 100 percent Chemical Weapons Convention milestone for destruction of former production facilities
- **1QFY08:** Achieve 45 percent Chemical Weapons Convention milestone for destruction of stockpile chemical agent
- **1QFY08:** Complete closure at Aberdeen, MD
- **3QFY08:** Complete destruction of VX-filled rockets at Anniston, AL, and GB-filled rockets at Pine Bluff, AR

ACQUISITION PHASE

Concept & Technology Development | System Development & Demonstration | Production & Deployment | Operations & Support

Chemical Demilitarization

FOREIGN MILITARY SALES
None

CONTRACTORS
EG&G (Tooele, UT)
Washington Demilitarization Company (Umatilla, OR; Pine Bluff, AR)
Washington Group International (Anniston, AL)
Parsons Infrastructure & Technology (Newport, IN)
Bechtel Aberdeen (Edgewood, MD)

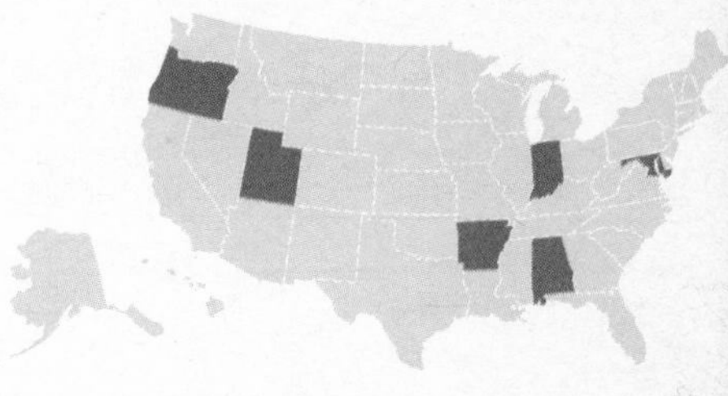

Chinook/CH-47F Improved Cargo Helicopter (ICH)

Conducts air assault, air movement, mass casualty evacuation, aerial recovery, and aerial resupply across the full spectrum of operations.

INVESTMENT COMPONENT

Modernization

Recapitalization

Maintenance

Description & Specifications

The Chinook/CH-47F Improved Combat Helicopter (ICH) upgrade program for the current CH-47D fleet will extend the service life of the current cargo helicopter fleet by an additional 20 years. The program includes a production line of new build aircraft and the remanufacture of all CH-47Ds in the current fleet to meet the total Chinook fielding requirement. Both new and remanufactured CH-47F ICHs incorporate new monolithic airframes and improvements to airframe reliability and maintainability. They provide an avionics architecture compliant with the DoD Information Technology Standards and Profile Registry (DISR), interoperability with DoD systems, and compliance with emerging Global Air Traffic Management (GATM) requirements.

Program Status

- **3QFY06:** CH-47F roll out
- **4QFY06:** Initiated key personnel training
- **1QFY07:** Award for Transportable Flight Proficiency Simulator production
- **1QFY07:** Delivery of first CH-47F new build aircraft
- **1QFY07:** Full rate production award

Recent and Projected Activities

- **2QFY07:** Complete initial operational test
- **4QFY07:** First unit equipped
- **1QFY08:** Multi-year procurement

ACQUISITION PHASE

Concept & Technology Development | System Development & Demonstration | **Production & Deployment** | Operations & Support

Chinook/CH-47F Improved Cargo Helicopter (ICH)

FOREIGN MILITARY SALES
None

CONTRACTORS
Aircraft:
Boeing (Philadelphia, PA)
Cockpit upgrade:
Rockwell Collins (Cedar Rapids, IA)
Engine upgrade:
Honeywell (Phoenix, AZ)
Extended range fuel system:
Robertson Aviation (Tempe, AZ)

MAX GROSS WEIGHT:	50,000 pounds
MAX CRUISE SPEED:	170 knots/184 miles per hour
TROOP CAPACITY:	36 (33 troops plus 3 crew members)
LITTER CAPACITY:	24
SLING-LOAD CAPACITY:	26,000 pounds center hook 17,000 pounds forward/aft hook 25,000 pounds tandem
MINIMUM CREW:	3 (pilot, co-pilot, and flight engineer)

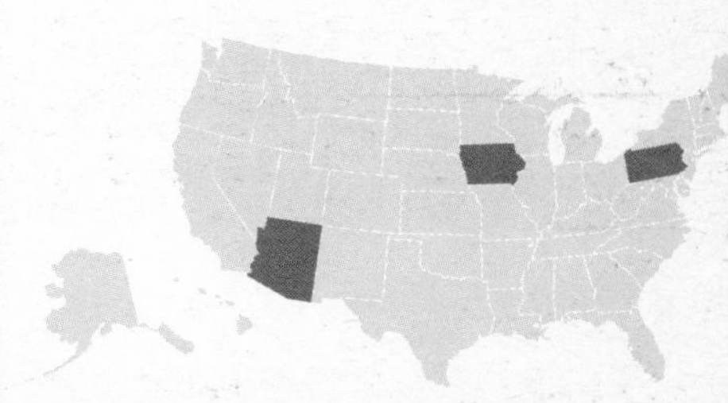

Close Combat Tactical Trainer (CCTT)

Provides training of armor, mechanized infantry, and cavalry units from platoon through battalion/squadron echelon, including staff, with a virtual, collective training simulator.

INVESTMENT COMPONENT

Modernization

Recapitalization

Maintenance

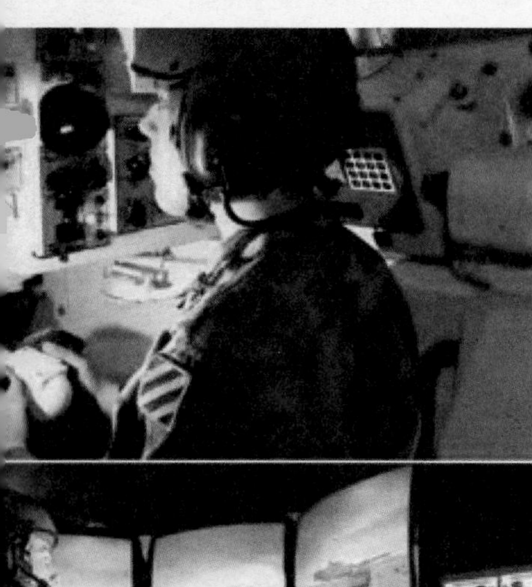

Description & Specifications

The Close Combat Tactical Trainer (CCTT) is a virtual, collective training simulator that is fully interoperable with the Aviation Combined Arms Tactical Trainer (AVCATT). Trainees operate from full-crew simulators and mock-up command posts. Crewed simulators—such as the Abrams Main Battle Tank family, the Bradley family, and the M113A3 armored personnel carrier—offer sufficient fidelity for collective mission accomplishment.

Infantry platoon and squad leaders can exit the Bradley Fighting Vehicle and move to dismounted infantry manned modules with control of virtual dismounted elements. Trainees use computer workstations in mock-up command posts to provide artillery, mortar, combat engineers, and logistics units to the synthetic battlefield.

Semi-automated forces workstations provide additional supporting units (such as aviation and air defense artillery) and all opposing forces. Thus, all battlefield operating systems are represented, ensuring effective simulation within a training environment that encompasses daylight, night, and fog conditions. CCTT supports training installations and posts in the United States, Europe, Korea, and Southwest Asia.

Program Status

- **3QFY06:** Mobile refurbishment of National Guard CCTT Louisiana Mechanized Platoon Set
- **3QFY06:** Fielding of CCTT XXI Force Battle Command Brigade and Below (FBCB2) at Ft. Hood
- **4QFY06:** Fielding of CCTT XXI FBCB2 Upgrades at Ft. Benning, Ft. Knox, and Ft. Stewart.
- **1QFY07:** Complete fielding of the initial Reconfigurable Vehicle Simulator to Ft. Hood, TX

Recent and Projected Activities

- **2QFY07:** Replace Fire Control System at all CCTT locations
- **3QFY07:** Production and fielding of the Reconfigurable Vehicle Simulator to CCTT installations
- **4QFY07:** Integration of Future Combat Systems (FCS) Spin Out 1 into CCTT at Ft. Bliss, TX
- **1QFY08:** Continue working with the FCS on capabilities required to train the Soldiers on FCS Spin Outs
- **1QFY08:** Continue production and fielding of the Reconfigurable Vehicle Simulator to CCTT installations. Begin production of Reconfigurable Vehicle Tactical Trainer to Infantry Brigade Combat Teams and other light units
- **FY08 and beyond:** Continue CCTT weapons systems currency and interoperability efforts and continue CCTT trainer unique performance improvement (technology refreshment) upgrades

ACQUISITION PHASE

Concept & Technology Development | System Development & Demonstration | Production & Deployment | Operations & Support

Operator Control

Dismounted Infantry

M2/M3

M1A2

M1A1

HMMWV

M113

After Action Review

CCTT SYSTEM CONFIGURATION

HIGH FIDELITY MANNED SIMULATORS

M1 (Variant) · M1A2 SEP · M113 · BFIST · HMMWV · M2 (Variant) · DSM'T INF · Interoperable Simulators

AFTER ACTION REVIEW

ATA
GGER

INITIALIZATION & MAINTENANCE

- MAINTENANCE CONSOLE (MC)
- MASTER CONTROL CONSOLE (MCC)

TACTICAL WORKSTATIONS

- MOTAR FIRE DIRECTION CENTER
- UNIT MAINTENANCE COMMAND POST
- COMBAT TRAIN (LOG) COMMAND POST
- TACTICAL AIR COMMAND POST
- FIELD ARTILLERY BN TOC
- COMBAT ENGINEER SUPPORT
- BRIGADE HQ
- BATTALION TF TOC

SEMI-AUTOMATED FORCES (SAF) WORKSTATIONS

ENEMY

FRIENDLY

TERRAIN DATABASES

esert (NTC)
emperate (Germany)
t Hood, TX
- Kosovo
- Korea

Close Combat Tactical Trainer (CCTT)

FOREIGN MILITARY SALES
None

CONTRACTORS
Lockheed Martin (Orlando, FL)
Rockwell Collins (Salt Lake City, UT)
Computer Sciences Corp. (San Diego, CA)
Kaegan Corp. (Orlando, FL)

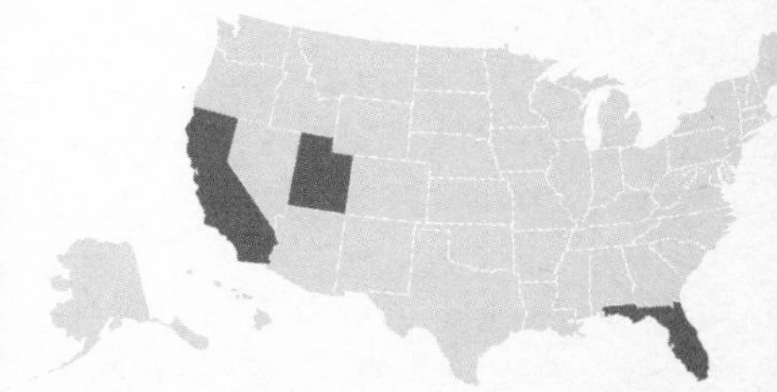

Combat Service Support Communications

Enhances Soldier safety and connects Soldiers with vital supplies through wireless satellite communications.

INVESTMENT COMPONENT

Modernization

Recapitalization

Maintenance

Description & Specifications

Combat Service Support Automated Information Systems Interface (CAISI) allows combat service support (CSS) automation devices to network and electronically exchange information via tactical or commercial communications. CSS satellite communications (CSS SATCOM) complements CAISI with an easy-to-use, transportable SATCOM link to extend broadband information exchange anywhere in the world.

CAISI is a deployable wireless LAN infrastructure linking up to 40 tents, vans, or shelters in a deployed 7-square-kilometer area. It includes Federal Information Processing Standards (FIPS) security requirements 140-2 Level 2 approved encryption for sensitive but unclassified traffic. CSS SATCOM uses commercial off-the-shelf auto-pointing remote satellite terminal hardware repackaged in fly-away transit cases. Four fixed-site, contractor-operated commercial teleports provide global coverage. CSS SATCOM supports operations at quick halt and rapid movement within the battle space, and eliminates the often dangerous transport of re-supply or spare parts orders on floppy disks.

CSS SATCOM uses commercial off-the-shelf, Ku band, fly-away remote satellite terminals called VSATs (very small aperture terminals) that are fielded to deployable forces, plus a contractor-operated supporting fixed infrastructure of four teleports and high-speed terrestrial links that is connected to the unclassified segment of the Global Information Grid.

Program Status

- **2Q–4QFY06:** CAISI and CSS VSAT trained and fielded equipment to units IAW the Army Campaign Plan
- CAISI upgrade (research and testing continues)
- Port of Beaumont, TX, networked using the new CAISI repeater modules
- CAISI deployed with CSS VSAT in support of Hurricane Katrina relief
- Migrated all CFLCC C4 VSATs into CSS SATCOM
- 554 VSATs fielded since FY04
- Biometric Identification System for Access supported with 12 VSATs
- **1FY07:** Commence fielding Voice over Internet Protocol (VoIP) with selected VSATs

Recent and Projected Activities

- **2QFY07–2QFY09:** CAISI and CSS VSAT will continue to train and equip units IAW the Army Campaign Plan; field units identified under G1 PSDR effort; begin "Tech Refresh"; CAISI accreditation for version 1.0 and 1.1
- **1QFY07–3QFY07:** Re-accredit CAISI and VSAT
- **1FY09:** Commence VSAT technology refresh

ACQUISITION PHASE

Concept & Technology Development | System Development & Demonstration | Production & Deployment | Operations & Support

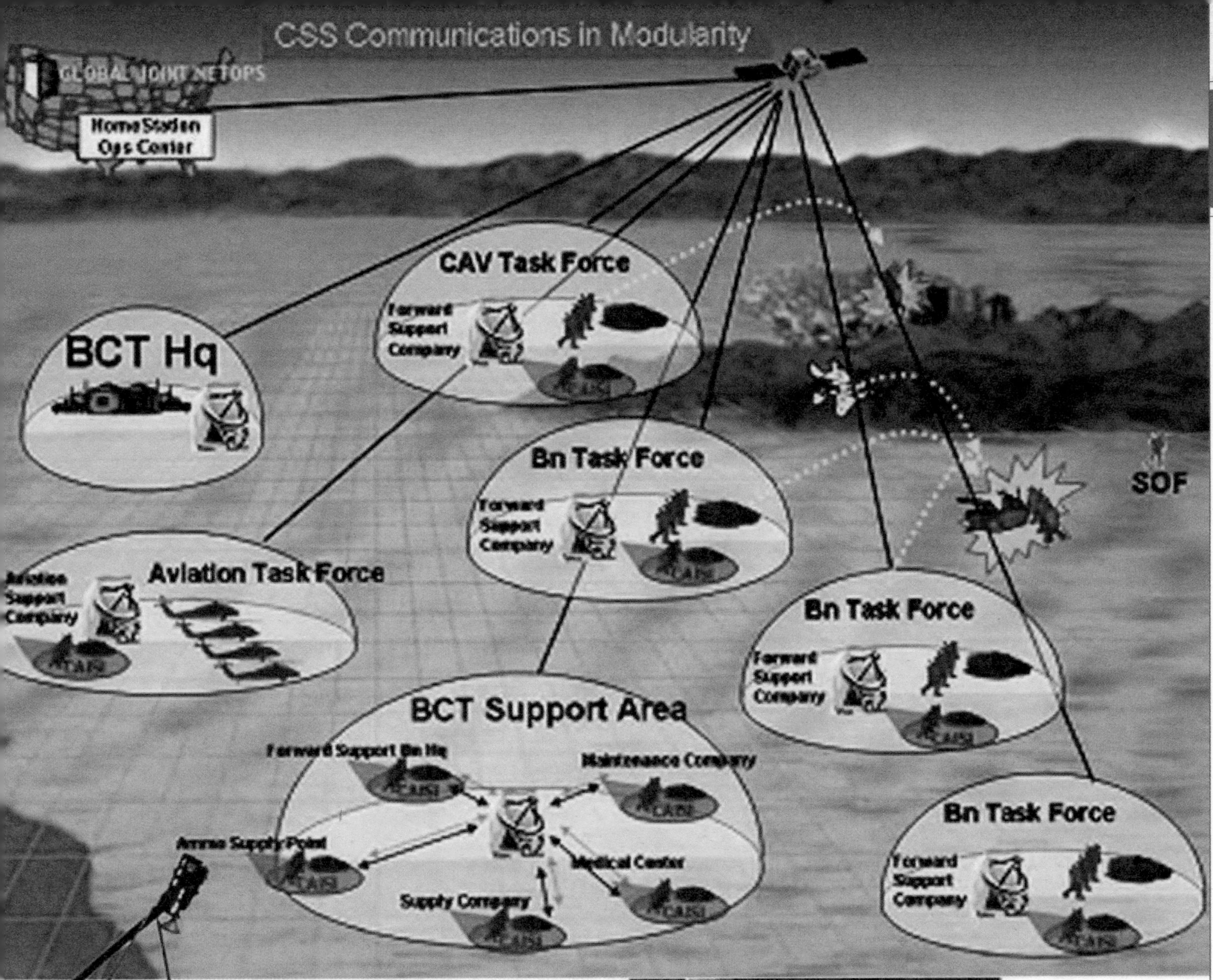

Combat Service Support Communications

FOREIGN MILITARY SALES
None

CONTRACTORS
IT equipment:
Computer Giant (New York, NY)
LTI DATACOM (Reston, VA)
APPTIS (Chantilly, VA)
GTSI (Chantilly, VA)
Project support/training:
Signal Solutions (Fairfax, VA)
TAMSCO (Ft. Monmouth NJ)
Satellite equipment:
Global Communications Solutions (Victor, NY)
Network infrastructure:
EyakTek (Anchorage, AK)
Segovia (Herndon, VA)

Common Hardware Systems (CHS)

Provides state-of-the-art tactical hardware and COTS software architectural building blocks for the warfighter that improves connectivity, interoperability, logistics, and worldwide maintenance on the C4ISR battlefield.

INVESTMENT COMPONENT

- Modernization
- Recapitalization
- Maintenance

Description & Specifications

The Common Hardware Systems (CHS) program provides state-of-the-art, fully qualified, interoperable, compatible, deployable, and survivable hardware and COTS software for command, control, and communications at all echelons of command for the United States Army and other DoD services. The CHS contract includes a technology insertion clause to continuously refresh the network-centric architectural building blocks, add new technology, and prevent hardware obsolescence. New products compliant with technology advances such as IPv6 can be easily added to the CHS offerings. They include a spectrum of computer processors from PDAs to high-end tactical computers, networking equipment, peripherals, displays, installation kits, and miscellaneous hardware needed for system integration. Three standardized environmental categories (V1, V2, and V3) are used to define hardware ruggedization and qualification test certification for the customers. V2 and V3 equipment items go through government-witnessed First Article Tests (FAT). Technical assistance and support services are also available. CHS currently has over 80 customers from Army and other DoD services.

CHS offers worldwide repair, maintenance, logistics, and technical support through strategically located contractor-operated CHS Repair Centers (CRCs) for tactical military units and management of a comprehensive 5-year warranty and 3-day turnaround for repairs.

CHS hardware Version 1 includes commercial workstations, peripherals, and networking products. Version 2 includes ruggedized workstations, peripherals, and networking products. Version 3 includes near-military specification rugged handheld units.

Program Status

- **3QFY05:** CHS-2 contract expired
- **1QFY06–4QFY06:** SWA contingency support
- **1QFY06–4QFY06:** CHS Out of Warranty Repair and Sustainment
- **2QFY06:** UID Implementation Plan signed
- **2QFY06:** 1st CAV Reset completed
- **3QFY06:** ABCE & Sustainment Support at NTC, Ft. Irwin and Ft. Hood, TX
- **4QFY06:** CHS-3 contract transferred from AMCOM Acq Ctr to CECOM Acq Ctr as a result of the PD CHS move from PM TOCs/AMDCCS to PM TRCS in FY06

Recent and Projected Activities

- **2QFY07–2QFY09:** CHS-3 hardware and software deliveries continue
- **2QFY07–2QFY09:** Support to the Global War on Terror continues (CHS-3 H/W and S/W; CHS SWA Repair Facility)
- **2QFY07–2QFY09:** Technology insertion to incorporate advances in computer technology and new user requirements and to replace technically obsolete (end of life) items continues
- **2QFY07:** CHS assumes distribution of SAASM GPS
- **2QFY07–4QFY07:** CHS/CERDEC Ipv6 lab experimentation continues
- **3QFY07–4QFY07:** Common testing of CHS-3 hardware continues

ACQUISITION PHASE

Concept & Technology Development | System Development & Demonstration | Production & Deployment | Operations & Support

Common Hardware Systems (CHS)

FOREIGN MILITARY SALES
None

CONTRACTORS
General Dynamics C4 Systems
(Taunton, MA)

SUBCONTRACTORS
Sun MicroSystems
Cisco
DRS
Hewlett Packard
TallaTech

TOC Hardware

V2 Processor

V2 Router

JNN Transit Cases

Iraq Repair Facility

CHS hardware includes:

- Super High Capacity Computer Unit (SHCU)
- High Capacity Computer Unit (HCU)
- Transportable Computer Unit (TCU)
- Standalone Computer Unit (SCU)
- Handheld Terminal Unit (HTU)
- Notebook Computer Unit (NCU)
- Four-Slice Multiple Processor Unit (4S MPU-2)
- AIS 3U CISC server
- Other high-end servers
- Laptops
- Tablets
- Routers
- Switches
- Firewalls
- Displays
- Peripherals
- Rugged Personal Digital Assistant (R-PDA)
- Uninterruptible Power Supplies and Conditioners

Conventional Ammunition Demilitarization

Performs end-of-life-cycle management for conventional ammunition, including disposition, demilitarization, and disposal with an emphasis on closed disposal.

INVESTMENT COMPONENT

Modernization

Recapitalization

Maintenance

Description & Specifications

Conventional Ammunition Demilitarization employs closed disposal technologies (resource recovery and recycling, reuse, and other processes) and open burn/ open detonation processes. The program uses government depots and commercial contractors within the United States and overseas, and supports military readiness as a supplier of necessary components in new production and maintenance operations. Critical explosives such as trinitrotoluene, or TNT, and tritonal, as well as components such as depleted-uranium penetrators and supplementary charges, are removed during demilitarization and reused in new production.

Funding for FY07 and FY08 provided for:

- Demilitarization through closed disposal technologies and open burn/ open detonation
- Army and DoD joint research and development of demilitarization technology
- Worldwide support requiring ammunition peculiar equipment
- Explosive safety efforts

Program Status

- **4QFY2006:** Demilitarization inventory (in short tons), 441,938; tons demilitarized, 30,338; tons received: 46,119
- **FY07:** Award of commercial contract option

Recent and Projected Activities

- **FY07:** Support for the joint U.S.-Korea Munitions Demilitarization Facility
- **Ongoing:** Demilitarization of persistent conventional mines as directed by presidential policy

ACQUISITION PHASE

Concept & Technology Development | System Development & Demonstration | Production & Deployment | Operations & Support

Conventional Ammunition Demilitarization

Open Burning

Open Detonation

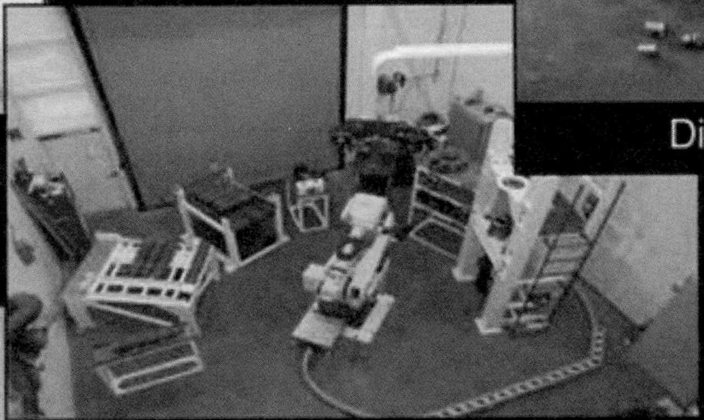

Disassembly

Robotic Disassembly of Projectiles (Sandia National Lab Project)

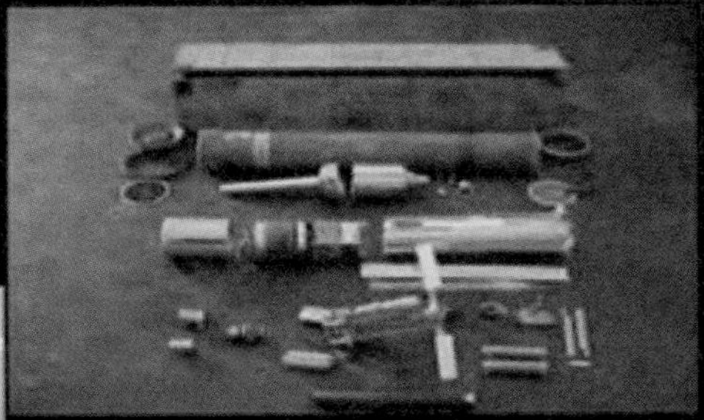

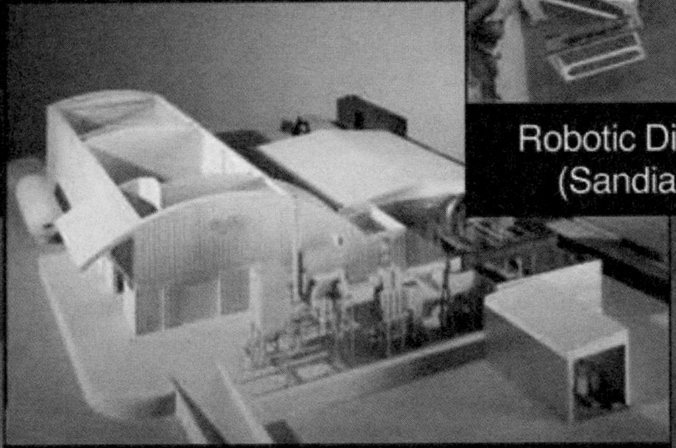

Cryofracture Demilitarization System (ARDEC Project, McAlester AAP)

Autoclave Meltout

FOREIGN MILITARY SALES
None

CONTRACTORS
General Dynamics Ordnance and Tactical Systems (St. Petersburg, FL)

Countermine

Provides Soldiers and maneuver commanders with a full range of countermine capabilities and immediate solutions to counter improvised explosive devices (IEDs).

INVESTMENT COMPONENT

Modernization

Recapitalization

Maintenance

Description & Specifications

The Countermine program seeks immediate solutions to the problem of improvised explosive devices (IEDs) being used against U.S. personnel in Operation Iraqi Freedom (OIF) and Operation Enduring Freedom (OEF). The program comprises several different systems:

- The Airborne Surveillance, Target Acquisition and Minefield Detection System (ASTAMIDS) is a Future Combat Systems (FCS) complementary program that detects minefields and obstacles from aerial platforms.
- The Ground Standoff Mine Detection System (GSTAMIDS) is an FCS complementary program that will provide mine detection, marking, and neutralization capabilities on Multifunction Utility/Logistics Equipment (MULE) countermine (CM) vehicles.
- The AN/PSS-14 Mine Detecting Set is a handheld multi-sensor mine detector being fielded Army-wide.
- The Area Mine Clearance System (AMCS) Foreign Comparative Test Program has two candidates: the Aardvark Mark IV and the Hydrema 910 mine clearing vehicles.
- The Vehicle Optics Sensor System (VOSS) will provide multi-sensor camera systems for route clearance and explosive ordnance disposal operations in Iraq and Afghanistan.
- Ferret Interrogation Arms for Route Clearance vehicles will provide greater capabilites for investigating suspected IEDs.

Program Status

- **2QFY05–2QFY06:** Fielded Buffalo Heavy Mine Protected Vehicles, RG-31 Medium Mine Protected Vehicles, and Interim Vehicle Mounted Mine Detectors for urgent operational needs in OIF and OEF
- **2QFY06:** AN/PSS-14 full rate production contract awarded
- **3QFY06:** ASTAMIDS preliminary design review
- **3QFY06:** GSTAMIDS FCS integrated baseline review
- **3QFY06:** Transitioned program management of Route Clearance vehicles to Project Manager Assured Mobility Systems
- **1QFY07:** ASTAMIDS component level critical design review
- **1QFY07:** AN/PSS-14 full rate production; Army-wide fielding continues
- **1QFY07:** Contract award for VOSS Phase 1 production contract

Recent and Projected Activities

- **2QFY07:** AMCS Foreign Comparative Test
- **3QFY07:** VOSS Phase 2 production contract award
- **3QFY07:** AMCS Milestone C
- **2QFY08:** GSTAMIDS FCS critical design review
- **2QFY09:** ASTAMIDS Milestone C and low rate initial production decision

ACQUISITION PHASE

Concept & Technology Development | System Development & Demonstration | Production & Deployment | Operations & Support

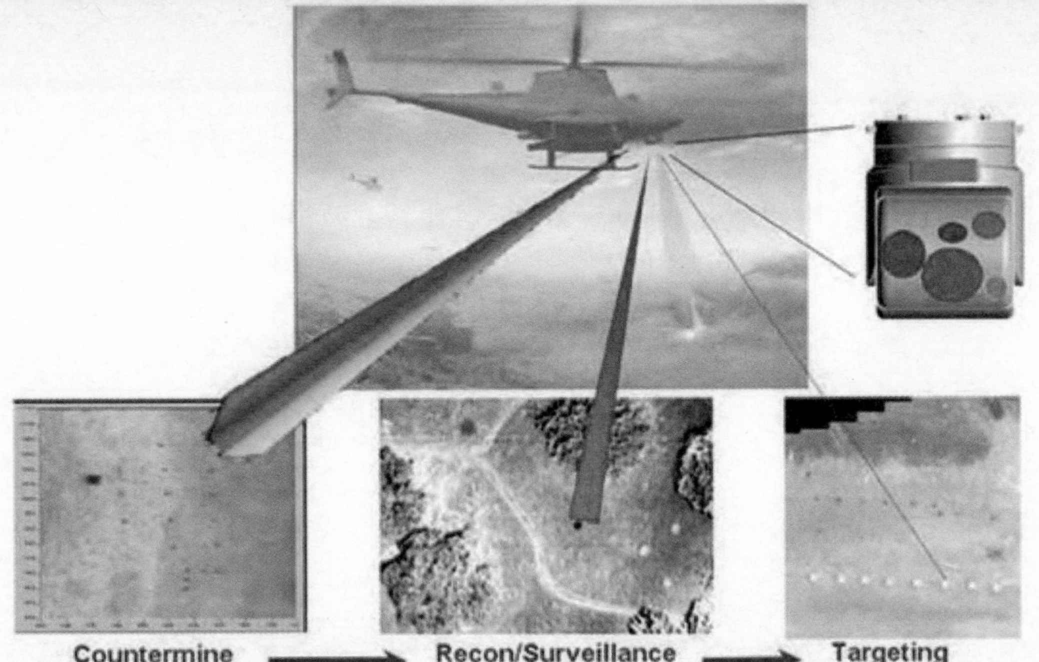

Countermine

FOREIGN MILITARY SALES
None

CONTRACTORS
AN/PSS-14:
CyTerra Corp. (Waltham, MA; Orlando, FL)
GSTAMIDS FCS:
BAE Systems (Austin, TX)
ASTAMIDS:
Northrop Grumman (Melbourne, FL)
VOSS:
Gyrocam Systems LLC (Sarasota, FL)

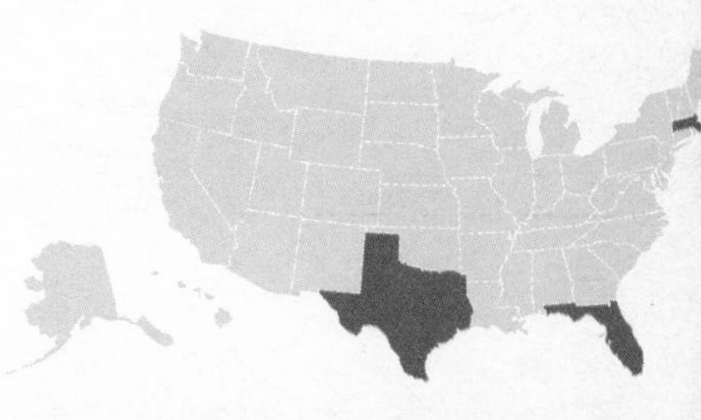

Defense Enterprise Wideband SATCOM System (DEWSS)

Provides secure, high-capacity voice and data communications and intelligence transfer to deployed forces worldwide.

INVESTMENT COMPONENT

Modernization

Recapitalization

Maintenance

Description & Specifications

The Defense Enterprise Wideband SATCOM System (DEWSS) provides super-high-frequency, beyond-line-of-sight communications, and is a critical conduit for intelligence information transfer. DEWSS consists of a geosynchronous orbiting satellite network, fixed enterprise military satellite terminals, baseband, payload control systems, and related equipment. DEWSS also provides reachback capability to sanctuary for deployed forces (teleport and standard tactical entry point sites). DEWSS modernization efforts provide tactical warfighters with reachback access to Defense Information Systems Network services, ensures survivable communications for critical nuclear command and control, and supports the Army's mission of payload and network control on super-high-frequency wideband communications satellites.

DEWSS is designed to satisfy long-term communication needs of warfighters and combatant commanders, as well as command, control, communications, and intelligence requirements. DEWSS provides the equipment that the U.S. Army Space and Missile Defense Command uses to perform its payload and network control mission on wideband satellites. DEWSS also provides an anti-jam and anti-scintillation capability for key strategic forces.

The DEWSS program includes modernization of enterprise terminals, baseband, and payload and network control systems required to support warfighter use of these satellites.

The DoD began launching Wideband Gapfiller Satellites (WGS) to provide warfighters with greatly increased capacity and new Ka-band capability in June 2007.

Program Status

- **3QFY06:** Awarded option for two Gapfiller Satellite Configuration Control Elements; completed Objective Defense Satellite Communications System Operations Control Workstation installs; modernization efforts to support WGS and Transformational Satellite continue; AN/GSC-52 modernization program continues to extend life for these terminals to 2015; installations, de-installations, and relocations of fixed strategic ground terminals and baseband continue as required by combatant commanders and validated by Joint Staff
- **4QFY06:** IMPCS Phase II CDR

Recent and Projected Activities

- **2QFY07:** Complete Direct Communications Link (DCL) Terminal B installation at Ft. Detrick, MD; Joint Management and Operations Subsystem (JMOS) contract award
- **4QFY07:** Complete KaSTARS terminal installations, complete Gapfiller Satellite Configuration Control Element (GSCCE) contractor On-Orbit Test; Common Network Planning Software (CNPS) government confidence test
- **2QFY08:** Award accelerated Modernization of Enterprise Terminals (MET) contract and RRFIS contract

ACQUISITION PHASE

Concept & Technology Development | System Development & Demonstration | Production & Deployment | Operations & Support

Defense Enterprise Wideband SATCOM System (DEWSS)

FOREIGN MILITARY SALES
None

CONTRACTORS
Satellite equipment:
ITT Industries (Colorado Springs, CO)
Boeing Satellite Systems (Los Angeles, CA)
Installation kits:
Harris Corp. (Melbourne, FL)
Software:
John Hopkins University Applied Physics Laboratory (Laurel, MD)
Northrop Grumman (Orlando, FL)
Engineering support:
U.S. Army Information Systems Engineering Command (Ft. Huachuca, AZ))

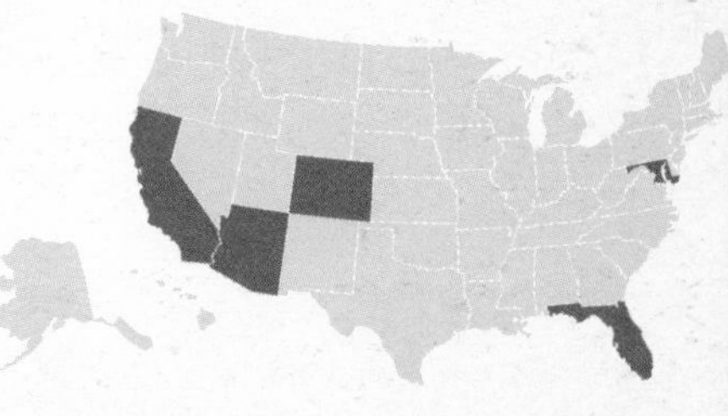

Distributed Common Ground System-Army (DCGS-A)

Provides commanders with threat, weather, and terrain data, as well as information and intelligence to synchronize the elements of joint and combined arms combat power.

INVESTMENT COMPONENT

- **Modernization**
- Recapitalization
- Maintenance

Description & Specifications

The Distributed Common Ground System-Army (DCGS-A) is the Army's primary system for tasking, processing, correlating, integrating, exploiting, and disseminating intelligence, surveillance, and reconnaissance (ISR) assets. DCGS-A assists in creating the common operational picture and enhances situational understanding, supporting the commander's ability to execute battle commands, synchronize fires and effects, rapidly shift battle focus, and protect the force. DCGS-A is the ISR gateway to joint, interagency, allied, coalition, and national data, information, intelligence, and collaboration.

DCGS-A emphasizes the use of reach and split-based operations to improve data access, reduce forward footprint, and increase interoperability via a network-enabled modular, tailorable system in fixed, mobile, and embedded configurations.

DCGS-A supports the ISR component embedded on Future Combat Systems (FCS) systems.

Program Status

- **2QFY06:** DCGS-A acquisition plan presented at Weapon System Review; Increment 1 approved
- **3QFY06:** Authorization to complete non-recurring engineering/design for mobile DCGS-A models at brigade and division/corps and integrate Version 3 baseline, the DCGS Integrated Backbone (DIB), and the JIOC-I Brain capabilities at the 513th fixed site
- **4QFY06:** Risk reduction assessment of DCGS-A V3.0a interoperability conducted at the Army's Central Technical Support Facility (CTSF), Ft. Hood, TX; DCGS-A V3.0b currently scheduled to begin assessment at CTSF mid-October 2006
- **4QFY06:** Contract awarded for design and integration of the DCGS-A mobile system (Version 4) to Northrop Grumman; system requirements review conducted; draft system specification submitted; mobile proof-of-concept hardware design demonstrated

Recent and Projected Activities

- **3QFY07:** Field DCGS-A Version 3 to Operation Iraqi Freedom and Operation Enduring Freedom
- **3QFY07:** DCGS-A participation in FCS 1.1 exercise
- **4QFY07:** DCGS-A V4 mobile first article

ACQUISITION PHASE

Concept & Technology Development | **System Development & Demonstration** | Production & Deployment | Operations & Support

Distributed Common Ground System-Army (DCGS-A)

FOREIGN MILITARY SALES

None

CONTRACTORS

V4 system integrator:

Northrop Grumman (Linthicum, MD)

Distributed Learning System (DLS)

Ensures that Soldiers receive critical mission training for mission success.

INVESTMENT COMPONENT

Modernization

Recapitalization

Maintenance

Description & Specifications

The Distributed Learning System (DLS) provides digital training facilities equipped with computers and video equipment enabling Soldiers to take digital training anywhere in the world at any time. Approximately 270 digital training facilities have been fielded worldwide.

DLS provides:

- Digital training facilities capable of delivering courseware for individual or group training
- Enterprise management of the DLS system
- A web-based learning management system for centralizing training management and delivery
- Deployed Digital Training Campuses (DDTC) (planned) to deliver multimedia courseware to deployed Soldiers
- Army e-learning: commercial web-based training for business, information technology, or language skills

Program Status

- **3QFY06–1QFY07:** Sustained a centrally managed global training enterprise; electronically delivered training in military occupational specialties; fielded Learning Management System to Army schools; completed Requirements Analysis for DDTC; increased Army e-Learning and Rosetta Stone enrollments; DLS named among top 50 Innovations in Government

Recent and Projected Activities

- **2QFY07:** Complete automated interface with legacy training systems
- **2QFY07:** Award DDTC contract
- **2QFY08:** Complete deployed digital training campus testing
- **2QFY0–2QFY09:** Continue fielding Army learning management system and DDTC; activate DLS disaster recovery site; sustain operation of DLS

ACQUISITION PHASE

Concept & Technology Development | System Development & Demonstration | Production & Deployment | Operations & Support

Distributed Learning System (DLS)

FOREIGN MILITARY SALES
None

CONTRACTORS
Army Learning Management System, Digital Training Facility, Enterprise Management Center Operations:
International Business Machines (Fairfax, VA)
Army e-learning:
Skillsoft (Nashua, NH)
Rosetta Stone language training services:
Fairfield (Harrisonburg, VA)
Program management support:
L3 Titan Corp. (San Diego, CA)

Dry Support Bridge (DSB)

Supports military load classification 100 (wheeled)/80 (tracked) vehicles over 40-meter gaps via a mobile, rapidly erected, modular military bridge.

INVESTMENT COMPONENT

- **Modernization**
- Recapitalization
- Maintenance

Description & Specifications

The Dry Support Bridge (DSB) system is fielded to Multi-Role Bridge Companies (MRBC) and requires a crew of eight Soldiers to deploy a 40-meter bridge in fewer than 90 minutes (daytime). The bridge modules are palletized onto seven flat racks and transported by equipment organic to the multi-role bridge company. DSB uses a launcher mounted on a dedicated Palletized Load System (PLS) chassis to deploy the modular bridge sections, which have a 4.3-meter road width and can span up to 40 meters. DSB is designed to replace the M3 Medium Girder Bridge.

Program Status

- **3QFY06:** Fielded to 200th Multi-Role Bridge Company
- **4QFY06:** Fielded to 502nd Engineer Company

Recent and Projected Activities

- **3QFY07:** Fielding to 652nd Engineer Company
- **FY07:** Begin design work on a floating DSB
- **FY08:** Begin production of a 46-meter gap capability
- **3QFY08** Fielding to 1437th Engineer Company
- **4QFY08:** Fielding to 739th Engineer Company
- **2QFY09:** Fielding to 671st Engineer Company
- **3QFY09:** Fielding to 1438th Engineer Company

ACQUISITION PHASE

- Concept & Technology Development
- System Development & Demonstration
- **Production & Deployment**
- Operations & Support

Dry Support Bridge (DSB)

FOREIGN MILITARY SALES
None

CONTRACTORS
Manufacturer:
Williams Fairey Engineering, Ltd. (Stockport, UK)
PLS chassis:
Oshkosh Truck (Oshkosh, WI)
Logistics:
XMCO (Warren, MI)

Engagement Skills Trainer (EST) 2000

Provides simulated indoor training for individual and crew-served weapons marksmanship, collective marksmanship, and "shoot-don't-shoot" training.

INVESTMENT COMPONENT

- Modernization
- Recapitalization
- Maintenance

Description & Specifications

The Engagement Skills Trainer (EST) 2000 is an indoor, multi-purpose, multi-lane, small arms, crew-served, and individual anti-tank simulator that saves ammunition resources and travel time and costs to and from ranges. It simulates weapon training events that lead to live-fire individual or crew weapon qualification and training events currently not resourced under the Standards in Training Commission (STRAC).

With the EST, squad leaders are able to control and evaluate individual, fire team, and squad performance. The EST 2000 simulates the following weapons: M16A2 rifle, M16A4, M4 carbine, M9 pistol, MK19 grenade machine gun, M249 squad automatic weapon, M240 machine gun, M136 (Anti-Tank 4), M1200 shotgun, M2 machine gun, and M203 grenade launcher. At the request of other programs, efforts are under way to model other weapon systems to be part of EST 2000: They include the XM8 carbine and the XM307/XM312 crew-served weapon.

Program Status

- **3QFY06:** Lot VII contract (48 subsystems) awarded
- **1QFY07:** Field five new escalation of force shoot-don't-shoot scenarios
- **1QFY07:** Complete fielding of Lot VI, 160 subsystems

Recent and Projected Activities

- **2QFY07:** Field 30 each Block I Upgrade to Production Lots I and II (new auto-tracker, brighter digital projector, more powerful Computer Processing Unit, Windows XP operating system) making them interoperable with Lots III–IV
- **2QFY07:** Award the Lot VIII contract (62 subsystems, including the following Preplanned Product Improvements upgrades: M16A2 magazine upgrade, heavy weapons for initial entry training suites, rifle/carbine X-leveling, and M16A2/M4 upgrade modification.
- **3QFY07:** Field Lot VII subsystems (48 subsystems)

ACQUISITION PHASE

- Concept & Technology Development
- System Development & Demonstration
- Production & Deployment
- Operations & Support

Engagement Skills Trainer (EST) 2000

FOREIGN MILITARY SALES
None

CONTRACTORS
Systems design, integration, and weapons:
Cubic Simulation Systems (formerly ECC International Corp.) (Orlando, FL)

Excalibur (XM982)

Provides improved fire support to the maneuver force with precision-guided, extended range, more lethal but collateral damage-reducing artillery projectiles.

INVESTMENT COMPONENT

Modernization

Recapitalization

Maintenance

Description & Specifications

The Excalibur (XM982) is a 155mm, Global Positioning System (GPS)-guided, fire-and-forget projectile under development as the Army's next-generation cannon artillery precision munition. The target, platform location, and GPS-specific data are entered into the projectile's mission computer through an enhanced portable inductive artillery fuze setter or automated system on the Future Combat Systems (FCS) Non-Line of Sight-Cannon (NLOS-C).

Excalibur uses a jam-resistant internal GPS receiver to update the inertial navigation system, providing precision guidance and dramatically improving accuracy regardless of range. Excalibur weighs 106 pounds, has three fuze options: height-of-burst, point-detonating, and delay/penetration; and is effective in all weather conditions and terrain.

The program is using an incremental approach to provide a combat capability to the Soldier as quickly as possible, and to deliver advanced capabilities and lower costs as technology matures. The initial variant (Block Ia-1) was fielded in 2007 to provide an urgently needed capability. It includes a unitary high-explosive warhead capable of penetrating urban structures, but is also effective against point targets, personnel targets, such as dismounted infantry and weapon crews, and light materiel targets, including air defense rockets, radars, and wheeled vehicles. Block Ia-2 will provide increased range (up to 40 kilometers) and reliability improvements. The third variant (Block Ib) will maintain performance and capabilities while reducing unit cost.

Excalibur is designed for fielding to the digitized Lightweight 155mm Howitzer, the 155mm M109A6 self-propelled howitzer (Paladin), the Future Force indirect fire weapon (FCS NLOS-C) and the Swedish Archer howitzer. Excalibur is an international cooperative program with Sweden, which contributes resources toward the development in accordance with an established project agreement and plans to join in procurement.

Program Status

- **3QFY06:** Completed developmental testing of Block Ia-1 and entered low rate initial production
- **1QFY07:** Completed first article testing and production verification testing of Block Ia-1

Recent and Projected Activities

- **3QFY07:** Field the first Excalibur precision artillery capability
- **3QFY07:** Milestone C: Enter the production and deployment phase for Block Ia-2
- **4QFY08:** Full materiel release and initial operational capability for Block Ia-2

ACQUISITION PHASE

Concept & Technology Development | System Development & Demonstration | Production & Deployment | Operations & Support

Excalibur (XM982)

FOREIGN MILITARY SALES
Canada (Block Ia-1)

CONTRACTORS

Systems integration:
Raytheon (Tucson, AZ)

Systems engineering:
BAE Systems Bofors Defense (Karlskoga, Sweden) teamed with Raytheon (Tucson, AZ)

Control actuator:
General Dynamics Ordnance and Tactical Systems (Healdsburg, CA)

GPS receiver:
L-3 Communications Interstate Electronics Corp. (Anaheim, CA)

Warhead:
General Dynamics Ordnance and Tactical Systems (Niceville, FL)

Extended Range Multi-Purpose (ERMP) Warrior Unmanned Aircraft System (UAS)

Provides combatant commanders a real-time responsive capability to conduct long-dwell, wide-area reconnaissance, surveillance, target acquisition, communications relay, and attack missions.

INVESTMENT COMPONENT

- **Modernization**
- Recapitalization
- Maintenance

Description & Specifications

The Extended Range Multi-Purpose (ERMP) Warrior Unmanned Aircraft System (UAS) addresses the need for a long-endurance armed unmanned aerial system that offers greater range, altitude, and payload flexibility.

The Warrior is powered by a heavy fuel engine (HFE) for higher performance, better fuel efficiency, and a longer lifetime. Its specifications include the following:

Length: 28 feet
Wingspan: 54 feet
Gross take off weight: 3,000+ pounds
Maximum speed: 130 kts.
Ceiling: 25,000 feet
Range: 300 kilometers
Endurance: 24 hours with a 250-pound payload

One ERMP Warrior UAS configuration, fielded in company sets, consists of:

- Twelve multi-role air vehicles (six with SATCOM)
- Five One Station Ground Control Stations
- Two Portable Ground Control Stations
- Five Tactical Common Data Link (TCDL) Ground Data Terminals
- Two TCDL Portable Ground Data Terminals
- One Ground SATCOM system
- Four Automatic Takeoff and Landing Systems
- Payloads: 12 Electro-Optical/ Infrared (EO/IR), 12 Synthetic Aperture Radar/Moving Target Indicator (SAR/MTI)
- Ground support equipment

Program Status

- **Current:** System development and demonstration

Recent and Projected Activities

- **3QFY08:** Limited user test
- **4QFY08:** Milestone C acquisition decision
- **4QFY09:** Initial operational test and evaluation
- **2QFY10:** Initial operational capability

ACQUISITION PHASE

- Concept & Technology Development
- **System Development & Demonstration**
- Production & Deployment
- Operations & Support

Extended Range Multi-Purpose (ERMP) Warrior Unmanned Aircraft System (UAS)

FOREIGN MILITARY SALES

None

CONTRACTORS

Air vehicle:
General Atomics (San Diego, CA)
Ground control station:
AAI (Hunt Valley, MD)
Tactical common data link:
L-3 Communications (Salt Lake City, UT)

Family of Medium Tactical Vehicles (FMTV)

Provides unit mobility/resupply, equipment/personnel transportation, and key ammunition distribution, using a family of vehicles based on a common chassis.

INVESTMENT COMPONENT

Modernization

Recapitalization

Maintenance

Description & Specifications

The Family of Medium Tactical Vehicles (FMTV) is a system of strategically deployable vehicles that perform general resupply, ammunition resupply, maintenance and recovery, and engineer support missions, and serve as weapon systems platforms for combat, combat support, and combat service support units in a tactical environment.

The Light Medium Tactical Vehicle (LMTV) has a 2.5-ton capacity (cargo and van models).

The Medium Tactical Vehicle (MTV) has a 5-ton capacity (cargo and long-wheelbase-cargo with and without materiel handling equipment, tractor, van, wrecker, and dump truck models). Three truck variants and two companion trailers, with the same cube and payload capacity as their prime movers, provide air drop capability. MTV also serves as the platform for the High Mobility Artillery Rocket System (HIMARS) and resupply vehicle for PATRIOT and HIMARS. MTV operates worldwide in all weather and terrain conditions.

FMTV enhances crew survivability through the use of hard cabs, three-point seat belts, and central tire inflation capability. FMTV enhances tactical mobility and is strategically deployable in C5, C17, C141, and C130 aircraft. It reduces the Army's logistical footprint by providing commonality of parts and components, reduced maintenance downtime, high reliability, and high operational readiness rate (more than 90 percent). FMTV incorporates a vehicle data bus and class V interactive electronic technical manual, significantly lowering operating and support costs compared with older trucks. Units are equipped with FMTVs at more than 68 locations worldwide, with more than 26,500 trucks and 4,200 trailers fielded as of September 8, 2006. The Army developed, tested, and installed add-on-armor and enhanced add-on-armor kits, and a Low Signature Armored Cab for Southwest Asia. Approximately 4,000 armored FMTVs have been deployed to Southwest Asia in support of Operation Iraqi Freedom.

Program Status

- **Current:** Full production and fielding to support Army transformation
- **1QFY07:** 10-ton dump truck production verification test start

Recent and Projected Activities

- Continue full production and fielding to support Army transformation
- **3QFY07:** 10-ton dump truck production verification test complete
- **4QFY07:** 10-ton dump truck limited user testing

ACQUISITION PHASE

Concept & Technology Development | System Development & Demonstration | Production & Deployment | Operations & Support

Family of Medium Tactical Vehicles (FMTV)

FOREIGN MILITARY SALES
Jordan

CONTRACTORS
Armor Holdings TVS (Sealy, TX)
Caterpillar (Greenville, SC)
Allison (Indianapolis, IN)
Arvin/Meritor (Newark, OH)
Precise Industries (Lufkin, TX)

	LMTV A1 Cargo	MTV A1 Cargo
Payload:	5,000 pounds	10,000 pounds
Towed load:	12,000 pounds	21,000 pounds
Engine:	Caterpillar 6-cylinder diesel	Caterpillar 6-cylinder diesel
Transmission:	Allison Automatic	Allison Automatic
Horsepower:	275	330
Drive:	4 x 4	6 x 6

FMTV A1 with Armor Kit

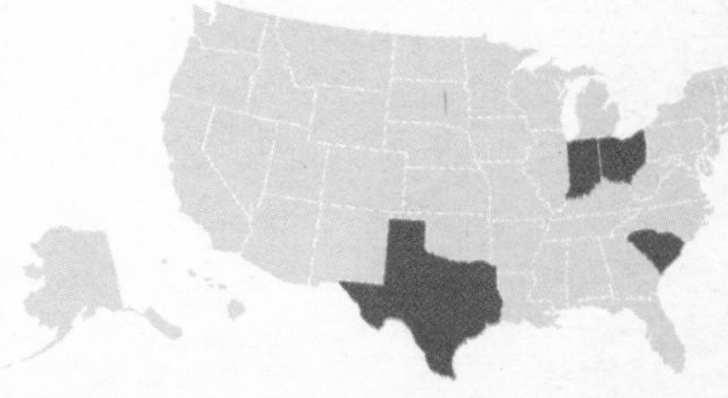

Fixed Wing

Performs operational support/focused logistics missions for the Army, joint services, national agencies, and multinational users in support of intelligence and electronic warfare, transporting key personnel, and providing logistical support for battle missions and homeland security.

INVESTMENT COMPONENT

- Modernization
- Recapitalization
- Maintenance

UC-35

Description & Specifications

The Fixed Wing fleet consists of eight aircraft platforms and 256 aircraft that allow the Army to perform day-to-day operations in a more timely and cost efficient manner without reliance on commercial transportation. The fleet provides timely movement of key personnel to critical locations throughout the theater of operations, and transports time-sensitive and mission critical supply items and repair parts needed to continue the war fight. Special electronic-mission aircraft (SEMA) provide commanders with critical intelligence and targeting information, enhancing lethality and survivability on the battlefield.

All Army fixed-wing aircraft are commercial off-the-shelf products or are developed from those products. The fleet includes:

- C-12 Utility
- C-20/C-37 Long range transport
- C-23 Cargo
- C-26 Utility
- EO-5 Airborne Reconnaissance Low (ARL)
- RC-12 Guardrail Common Sensor (GR/CS)
- UC-35 Utility

The EO-5 and RC-12 are classified as special electronic mission aircraft and provide real-time intelligence collection in peace and wartime environments. The C-12, C-23, C-26, and UC-35 are classified as operational support aircraft and provide direct fixed-wing support to warfighting combatant commanders worldwide. The C-20 and C-37 are assigned to Andrews Air Force Base and are classified as senior support aircraft for the chief of staff and service secretary.

Program Status

- C-12, RC-12, and UC-35 aircraft are sustained using a Life Cycle Contractor Support (LCCS) maintenance contract (DynCorp)
- C-23 aircraft are sustained using an LCCS maintenance contract (M7 Aerospace)
- C-37 and UC-35 were all purchased with Congressional plus-up funding
- C-37, C-20, and C-26 aircraft are sustained using Air Force LCCS maintenance contracts (Gulfstream and M7 Aerospace)
- EO-5 aircraft are sustained using an LCCS maintenance contract (King Aerospace)

Recent and Projected Activities

- Acquire three aircraft for the Army Parachute Team (Golden Knights)

ACQUISITION PHASE

- Concept & Technology Development
- System Development & Demonstration
- Production & Deployment
- Operations & Support

C-20/37

C-26

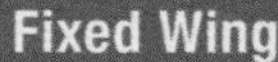

RC-12

EO-5 (ARL)

C-12

C23

	EO-5	C-12/RC-12	C-20/37	C-23	C-26	UC-35
Platform:	DeHavilland Dash 7	Beech King Air 200	Gulfstream GIV and GV	Shorts Sherpa	Fairchild Metro Liner	Cessna Citation
Propulsion:	PT6A-50	PT6A-41/42/67	RR 611-8/BR 710-48	PT6A-65AR	Garrett TPE331-12	JT15D or PW535A
Ceiling (feet):	25,000	31,000/35,000	45,000	20,000	25,000	45,000
Speed (knots):	110 (loiter) 220 (Cruise)	260	459	180	260	415
Max. Weight (pounds):	47,000	12,500/16,500	74,600/95,000	25,600	16,500	16,500
Range (nautical miles):	1,500	1,454/1,000	4,220/5,500	900	1,500	1,500
Passengers:	N/A	6-8/(N/A)	12-14	30	20	8

Fixed Wing

FOREIGN MILITARY SALES
None

CONTRACTORS
DynCorp (Ft. Worth, TX)
Gulfstream (Savannah, GA)
King Aerospace (Addison, TX)
M7 Aerospace (San Antonio, TX)

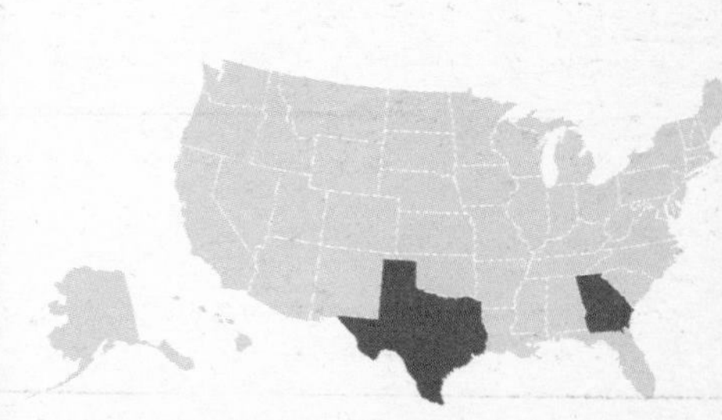

Force XXI Battle Command Brigade-and-Below (FBCB2)

Provides enhanced situational awareness to the lowest tactical level—the individual Soldier—and a seamless flow of command and control information across the battlefield.

INVESTMENT COMPONENT

- Modernization
- Recapitalization
- Maintenance

Description & Specifications

The Force XXI Battle Command Brigade-and-Below (FBCB2) forms the principal digital command and control system for the Army at brigade levels and below. It provides increased situational awareness on the battlefield by automatically disseminating throughout the network timely friendly force locations, reported enemy locations, and graphics to visualize the commander's intent and scheme of maneuver.

FBCB2 is a key component of the Army Battle Command System (ABCS). Applique hardware and software are integrated into the various platforms at brigade and below, as well as at appropriate division and corps slices necessary to support brigade operations. The system features the interconnection of platforms through two communication systems: FBCB2-Enhanced Position Location Reporting System (EPLRS) supported by the Tactical Internet and FBCB2-Blue Force Tracking supported by L-band satellite.

Program Status

- **3QFY06:** Awarded five-year full rate production contract
- **3QFY06:** Started fielding Stryker Brigade Combat Teams (SBCTs) 5 and 6, 3rd Infantry Division, and Armored Security Vehicles (ASVs)
- **4QFY06:** Started fielding 82nd Airborne (ABD) and continued fielding SBCTs 5 and 6, 3rd ID, and ASVs
- **1QFY07:** Complete fielding 82nd ABD and continue fielding SBCTs 5 and 6, 3rd ID, and ASVs

Recent and Projected Activities

- **2QFY07:** Start fielding 2nd ID Headquarters Heavy Brigade Combat Team (HQ HBCT) 1; complete fielding SBCT 5 and 3rd ID and continue fielding ASVs and SBCT 6
- **3QFY07:** Start fielding 30 HBCT and continue fielding ASVs and SBCT 6
- **4QFY07:** Complete fielding ASVs, SBCT 6, and 30 HBCT
- **4QFY07:** Follow-on test and evaluation
- **2QFY08:** Start fielding FBCB2 software V6.5 to Operation Iraqi Freedom (OIF)
- **1QFY09:** Fielding joint capabilities release software V1.0 to OIF

ACQUISITION PHASE

- Concept & Technology Development
- System Development & Demonstration
- Production & Deployment
- Operations & Support

Force XXI Battle Command Brigade-and-Below (FBCB2)

FOREIGN MILITARY SALES
Australia

CONTRACTORS
Software/systems engineering:
Northrop Grumman Space & Mission Systems Corp. (Redondo Beach, CA)
Hardware: DRS Technologies (Palm Bay, FL)
Installation kits: Northrop Grumman Space & Mission Systems Corp. (Redondo Beach, CA)
Satellite services:
COMTECH (Germantown, MD)
Field support:
Engineering Solutions & Products, Inc. (Eatontown, NJ)

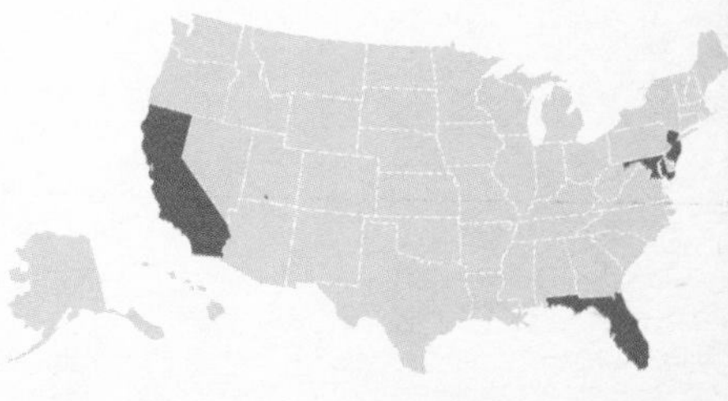

Forward Area Air Defense Command and Control (FAAD C2)

Collects, digitally processes, and disseminates real-time target tracking and cuing information, common tactical air picture, and command, control, and intelligence information to all short-range air defense weapons.

INVESTMENT COMPONENT

- **Modernization**
- Recapitalization
- Maintenance

Description & Specifications

Forward Area Air Defense Command and Control (FAAD C2) is a battle management and command, control, communications, computers, and intelligence system. Unique FAAD C2 software provides mission capability, such as critical C2, situational awareness, and automated air track information, by integrating engagement operations software for multiple systems, including:

- Avenger
- Bradley Stinger Fighting Vehicle
- Sentinel
- Army Battle Command System (ABCS)

FAAD C2 supports air defense weapon systems engagement operations by tracking friendly and enemy aircraft, cruise missiles, and unmanned aerial vehicles, and linking to other Army battle command systems and performing C2 for Avenger and Bradley Linebacker. It also performs C2 for the C-RAM system. FAAD C2 uses the following communication systems:

- Enhanced Position Location Reporting System (EPLRS)
- Joint Tactical Information Distribution System (JTIDS)
- Single Channel Ground and Airborne Radio System (SINCGARS)

FAAD C2 provides joint C2 interoperability and horizontal integration with all other Army C2 and air defense artillery systems, including, but not limited to:

- Surface Launched Advanced Medium Range Air-to-Air Missile (SLAMRAAM)
- PATRIOT
- Theater High Altitude Area Defense (THAAD)
- Airborne Warning and Control System (AWACS)
- ABCS

Program Status

- **1–2QFY06:** Field ADAM-5 to 2/25 Infantry Division SBCT 5
- **4QFY06:** C-RAM testing
- **1QFY07:** Limited user test

Recent and Projected Activities

- **2QFY07-C-RAM:** Live fire demonstration
- **3QFY07:** FAAD participation in Joint Red Flag exercise
- **1QFY08:** C-RAM sense and warn forward operating base fieldings completed
- **2QFY08:** FAAD Version 5.4B materiel release
- **2QFY08–2QFY09:** FAAD Version 5.4B scheduled software upgrade
- **3QFY08:** FAAD fielded to final National Guard unit

ACQUISITION PHASE

Concept & Technology Development | System Development & Demonstration | **Production & Deployment** | Operations & Support

Forward Area Air Defense Command and Control (FAAD C2)

FOREIGN MILITARY SALES
Sold to Egypt, first case in 3QFY02

CONTRACTORS

Software:
Northrop Grumman Space & Mission Systems Corp.
(Redondo Beach, CA)

Hardware:
Northrop Grumman (Huntsville, AL)

SETA:
CAS, Inc. (Huntsville, AL)

CHS 3 :
General Dynamics (Taunton, MA)

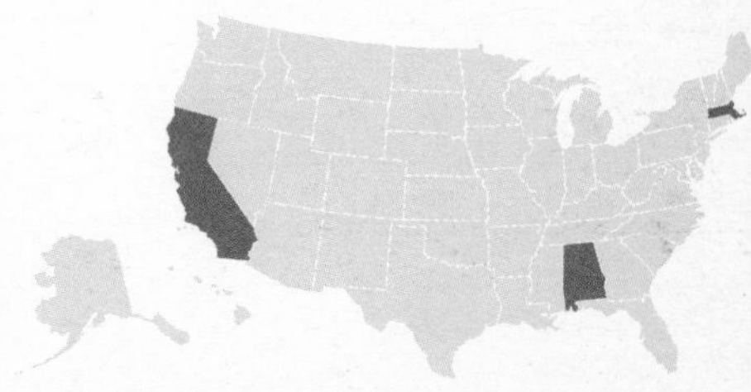

Forward Repair System (FRS)

Repairs battle-damaged combat systems "on-site," up through the direct support level, in the forward battle area.

INVESTMENT COMPONENT

Modernization

Recapitalization

Maintenance

Description & Specifications

The Forward Repair System (FRS) is a high-mobility, forward-maintenance/repair module system that reduces man-hours for maintenance personnel. Mounted to a flatrack, it is transported by Palletized Load System (PLS) trucks in Force XXI Divisions, or by Heavy Expanded Mobility Tactical Truck-Load Handling System (HEMTT-LHS) in Stryker Brigade Combat Teams (SBCTs).

- Dimensions: 8 feet wide by 8 feet high by 20 feet long
- Weight: 24,600 pounds
- Air transportability: C-130, C-141
- Crane capacity: Remove and replace major components from supported tracked and wheeled vehicles. Crane is capable of lifting and maneuvering 10,000 pounds at 14-foot radius. Frees recovery assets to perform their intended mission.
- Generator capacity: Provides 35 kilowatts at 60Hz output. Capable of simultaneously providing electrical power for all subsystems, including power take-off (PTO) for crane. Noise level is 73dB at 7 meters.
- Air compressor: Provides 175 pounds per square inch at 50 cubic feet per minute. Eighty-gallon capacity supplies air for on-board pneumatic tools, inflation of tires, and compressed air cleaning.
- Welding and cutting equipment: Limited spot welding and cutting equipment (shielded metal and "stick" welding, metal inert gas "MIG" welding, and exothermic cutting/brazing).
- Tools: Industrial grade hand/pneumatic/power tools. FRS's tool load is functionally equivalent to #1 common tool kit
- Shelter/protection: Contains a canvas tarp and heater that protects from the weather yet preserves access to welding, air, and accessory tools
- Air jacks/bags: Two, Kevlar, each capable of lifting 40,000 pounds up to 15 inches
- Rapid inventory: 720 line items are stored in draws with foam cut-outs of each tool for rapid inventory identification.

Program Status

- **Current:** In production

Recent and Projected Activities

- Field to 1 MX AV BDE; 577th Engineering Battalion (TRADOC); SBCT 6; 3rd Infantry Division, 3rd and 4th Brigade Combat Team; 30 AR BDE; North Carolina National Guard; 1st Armored Division, 12th Combined Arms Battalion

ACQUISITION PHASE

Concept & Technology Development | System Development & Demonstration | Production & Deployment | Operations & Support

Forward Repair System (FRS)

FOREIGN MILITARY SALES
None

CONTRACTORS
Rock Island Arsenal (Rock Island, IL)
Grove Worldwide (Shady Grove, PA)
Snap-on Industrial (Crystal Lake, IL)
Cummins Power (Minneapolis, MN)
Ingersoll-Rand (Campbellsville, KY)

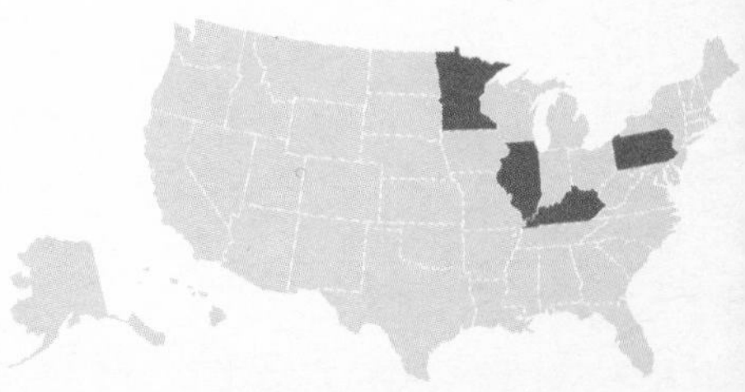

Global Combat Support System-Army (GCSS-Army)

Enables users to perform logistics support tasks and to view complete logistics management information and make timely, data-driven decisions.

INVESTMENT COMPONENT

Modernization

Recapitalization

Maintenance

Description & Specifications

The Global Combat Support System-Army (GCSS-Army) will replace 13 legacy Army logistics systems and interface or integrate with applicable command and control (C2) and joint systems to enhance combat support/combat service support (CS/CSS) transformation. GCSS-Army provides the primary logistics system enabler to achieve the Army transformation vision of a technologically advanced, enterprise resource planning (ERP) system, capable of managing the flow of the CS/CSS logistics resources and information, to satisfy the Army's logistics modernization requirements.

GCSS-Army provides commanders with the capability to anticipate, allocate, and synchronize the flow of CSS resources to equip, deploy, and project, sustain, reconstitute, and re-deploy forces in support of the national military strategy, providing rapid, coordinated, and sustained CSS to the Army, joint services, and allied forces within a reduced footprint.

The web-based system, supported by lightweight mobile applications, provides essential functionality for limited disconnected operation, and robust deployable communications capable of providing reachback to a centralized data repository regardless of location—e.g., sustaining base or deployed theater—for all users at all echelons.

GCSS-Army will meet the needs of the warfighter by re-engineering field Army logistics business processes to provide the right materiel at the right time while anticipating warfighter requirements for asset visibility and control, along with timely and accurate management information. GCSS-Army is the field Army (tactical) component of the Army's single integrated logistics solution capable of supporting rapid force projection and battlefield functional areas of manning, arming, fixing, fueling, moving, and sustaining deployed forces.

Program Status

- **2QFY05:** Contract awarded for Product Lifecycle Management Plus (PLM+)
- **1QFY06:** Acquisition strategy revised to accommodate existing funding for FY06-07
- **1QFY07:** Milestone B decision, enter realization phase

Recent and Projected Activities

- **4QFY07:** Complete realization phase including configuration of software, unit testing and integration testing; conduct operational assessment for supply functionality above unit level at a single unit
- **1QFY08:** Award contract for integration of maintenance, property book, and ammunition functionality and configure system
- **FY09:** Conduct Operational Assessment for maintenance, property book, and ammunition functionality at a single unit.

ACQUISITION PHASE

Concept & Technology Development | System Development & Demonstration | Production & Deployment | Operations & Support

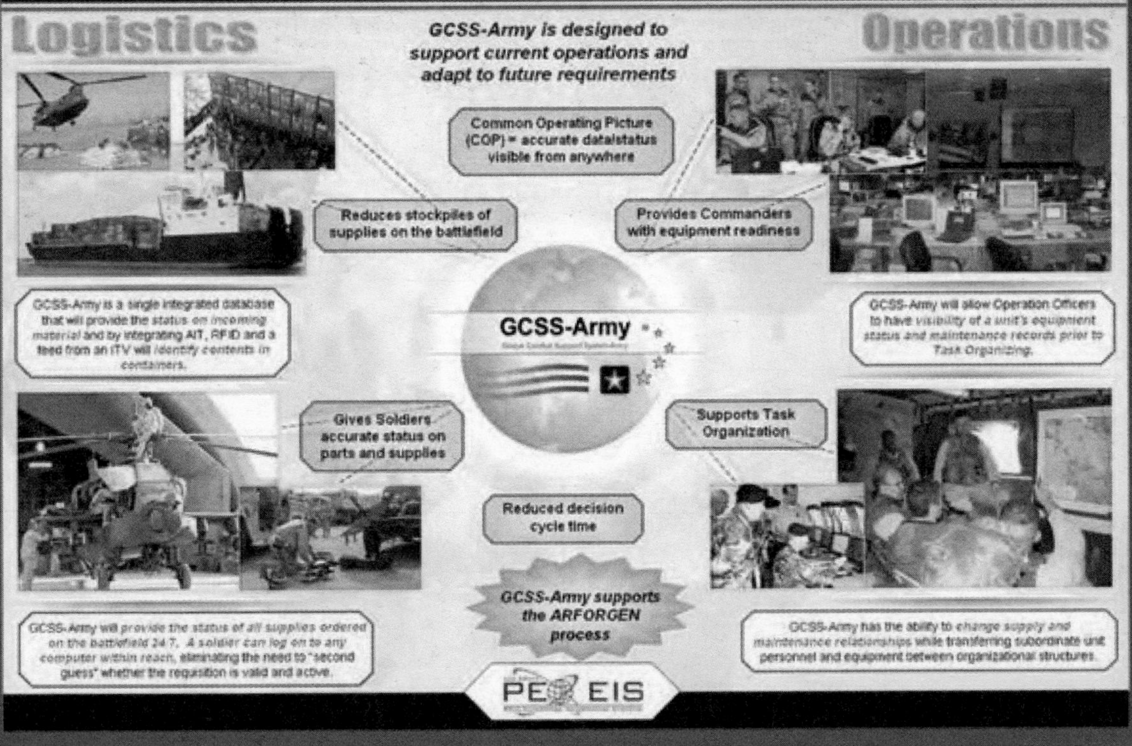

A critical component of the Single Army Logistics Enterprise (SALE)

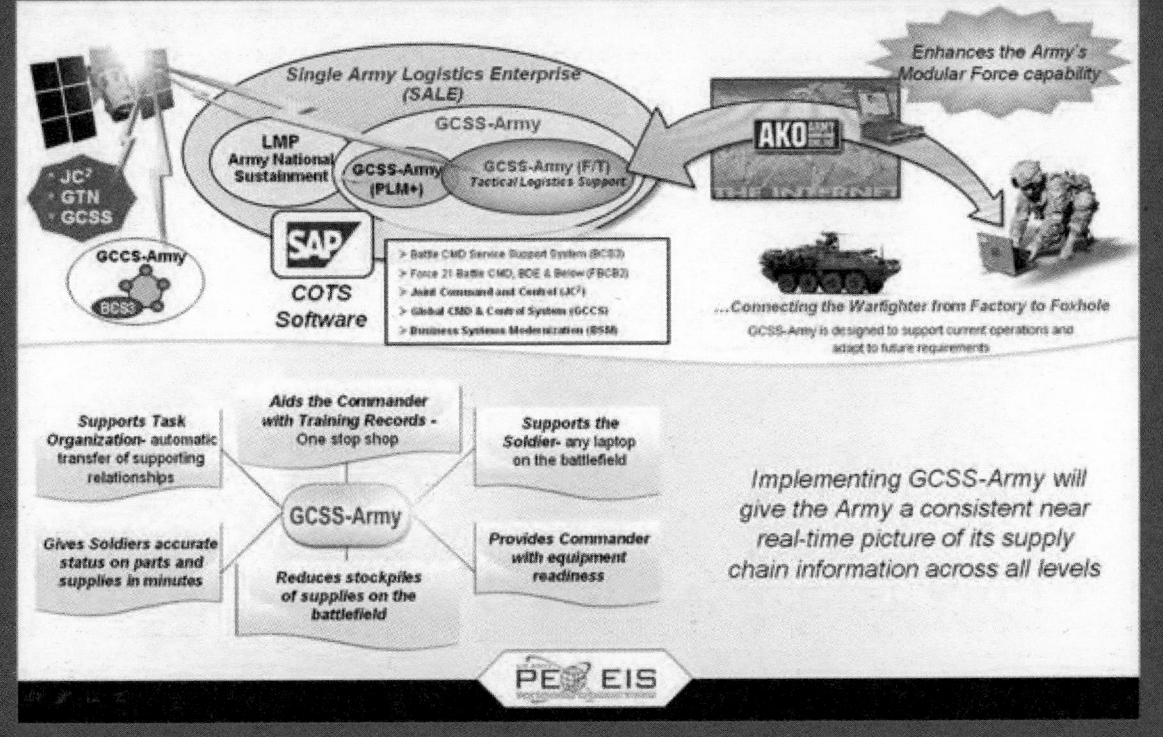

Global Combat Support System-Army (GCSS-Army)

FOREIGN MILITARY SALES
None

CONTRACTORS

Lead system integrator:
Northrop Grumman Mission Systems (Chester, VA)

Program support:
L-3 Communications (Colonial Heights, VA)

Global Command and Control System-Army (GCCS-A)

Enhances warfighter capabilities across the spectrum of conflict, during joint and combined operations, through automated command and control tools for strategic and operational commanders.

INVESTMENT COMPONENT

- **Modernization**
- Recapitalization
- Maintenance

Description & Specifications

The Global Command and Control System-Army (GCCS-A) is the Army's strategic and operational command and control (C2) system, providing readiness, planning, mobilization, and deployment capability information for strategic commanders. For theater commanders, GCCS-A provides the following:

- Common operational picture and associated friendly and enemy status information
- Force-employment planning and execution tools (receipt of forces, intra-theater planning, readiness, force tracking, onward movement, and execution status)
- Overall interoperability with joint, coalition, and the tactical Army Battle Command System (ABCS)

GCCS-A supports Army units from the strategic commanders and regional combatant commanders in the theater, down through the joint task force commander. As part of ABCS, GCCS-A provides a seamless Army extension from the joint GCCS system to echelons corps and below. Compatibility and interoperability are achieved by building the GCCS-A applications to function on the common operating environment and through interfaces with other C2 systems within the Army and other services.

The common operating environment specifies a common system infrastructure for all C2 systems in accordance with the joint technical architecture guidelines, which provide common support architecture and modular software for use by the services and agencies in developing mission-specific solutions to their C2 requirements. The hardware platform is based on commercial, off-the-shelf hardware. The system users are linked via local area networks in client/server configurations with an interface to the Secret Internet Protocol Router Network for worldwide communication.

Program Status

- **2QFY05–1QFY07:** Continue integration and testing of GCCS-A Block IV
- **2QFY05–1QFY07:** Support Operations Enduring Freedom and Iraqi Freedom (OEF/OIF)
- **2QFY05–1QFY07:** Support Network Enabled Commanders Capability (NECC) Milestone A & B efforts to include development of the capabilities description document with the Army annex with unique situational understanding requirements
- **2QFY05–1QFY07:** Develop and field Defense Readiness Reporting System -Army (DRRS-A) Force Readiness Tool (Phase 3)

Recent and Projected Activities

- **2QFY07–4QFY07:** Continue integration and testing of GCCS-A Block IV Releases 4.1 and 4.2
- **2QFY07–1QFY08:** Continue spiral development in support of DRRS-A Phase 4 requirements
- **2QFY07–1QFY09:** Continue directed fieldings and required support for OEF/OIF
- **2QFY07–1QFY09:** Continue support to NECC technology demonstration phase and transfer of GCCS-A capability to NECC

ACQUISITION PHASE

Concept & Technology Development | System Development & Demonstration | **Production & Deployment** | Operations & Support

Global Command & Control System - Army

Global Command and Control System-Army (GCCS-A)

FOREIGN MILITARY SALES
None

CONTRACTORS

Prime:
Lockheed Martin (Springfield, VA)

Software:
Lockheed Martin (Springfield, VA)
GESTALT (Camden, NJ)

Hardware:
General Dynamics (Taunton, MA)
GTSI Corp. (Chantilly, VA)

Fielding support:
FC Business Systems (Springfield, VA)
Engineering Solutions & Products (Eatontown, NJ)

Software training:
FC Business Systems (Atlanta, GA)

Global Positioning System (GPS)

Provides real-time position, velocity, and timing data to tactical and strategic organizations.

INVESTMENT COMPONENT

- **Modernization**
- Recapitalization
- Maintenance

Description & Specifications

The Global Positioning System (GPS) is a space-based joint service navigation program, led by the Air Force, which distributes position, velocity, and timing (PVT) data. It has three segments: a space segment (nominally 24 satellites), a ground control segment, and a user equipment segment. User equipment consists of receivers configured for handheld use, ground, aircraft, and watercraft applications. Military GPS receivers use the Precise Positioning Service (PPS) signal to gain enhanced accuracy and signal protection not available to commercial equipment. GPS receivers in the Army today are the Precision Lightweight GPS Receiver (PLGR) with more than 100,000 in handheld, installed, and integrated applications and the Defense Advanced GPS Receiver (DAGR) with more than 47,000 fielded. In addition, GPS user equipment includes a Ground-Based GPS Receiver Applications Module (GB-GRAM).

Over 18,000 GB-GRAMs have been procured and provide an embedded PPS capability to a variety of weapon systems. The Army represents more than 80 percent of the requirement for user equipment.

DAGR
Size: 6.37 x 3.4 x 1.56 inches
Weight: One pound; fits in a two-clip carrying case that attaches to Load Bearing Equipment
Frequency: Dual (L1/L2)
Battery Life: 19 hours (4 AA batteries)
Security: Selective availability anti-spoofing module
Satellites: All-in-view

GB-GRAM
Size: 0.6 x 2.45 x 3.4 inches
Weight: 3.5 ounces
Frequency: Dual (L1/L2)
Security: Selective availability anti-spoofing module
Satellites: All-in-view

Program Status

- **3QFY06–1QFY07:** Continue DAGR fieldings

Recent and Projected Activities

- **2QFY07–1QFY09:** Continue DAGR fieldings

ACQUISITION PHASE

- Concept & Technology Development
- System Development & Demonstration
- **Production & Deployment**
- Operations & Support

Global Positioning System (GPS)

FOREIGN MILITARY SALES
PPS-capable GPS receivers have been sold to 38 authorized countries

CONTRACTORS
DAGR/GB-GRAM acquisition and PLGR support:
Rockwell Collins (Cedar Rapids, IA)

Guardrail Common Sensor (GR/CS)

Provides signal intelligence collection and precision targeting that intercepts, collects, and precisely locates hostile communications intelligence radio frequency emitters and electronic intelligence threat radar emitters.

INVESTMENT COMPONENT

Modernization

Recapitalization

Maintenance

Description & Specifications

The Guardrail Common Sensor (GR/CS) is a corps-level, fixed-wing, airborne, signals intelligence (SIGINT) collection and precision targeting location system. It provides near-real-time information to tactical commanders in the corps/joint task force area with emphasis on deep battle and follow-on forces attack support. It collects low-, mid-, and high-band radio signals; identifies and classifies them; determines source location; and provides near-real-time reporting, ensuring information dominance to commanders. GR/CS uses an integrated processing facility (IPF), the control, data processing, and message center for the system. It includes:

- Integrated communications intelligence (COMINT) and electronic intelligence (ELINT) collection and reporting
- Enhanced signal classification and recognition and precision emitter geolocation
- Near-real-time direction finding
- Advanced integrated aircraft cockpit
- Tactical Satellite Remote Relay System (Systems 1, 2, and 4)

A standard system has eight to 12 RC-12 aircraft flying operational missions in sets of two or three. Up to three airborne relay facilities (ARFs) simultaneously collect communications and noncommunications emitter transmissions and gather lines of bearing (LOBs) and time difference of arrival (TDOA) data, which is transmitted to the IPF, correlated, and supplied to supported commands. One GR/CS system is authorized per aerial exploitation battalion in the military intelligence brigade at each corps.

Planned improvements include an enhanced precision geolocation subsystem, CHALS-C, with increased frequency coverage and a higher probability to collect targets, and a modern COMINT infrastructure and Core COMINT subsystem, Enhanced Situational Awareness (ESA), providing a frequency extension, a capability to process special signals, and elimination of non-supportable hardware and software. Ground processing software and hardware are being upgraded for interoperability with the DCGS-A architecture and Distributed Information Backbone.

Program Status

- **2QFY06:** System 2 adapter cable contract award
- **2QFY06:** X-Midas B-Kit contract award, 22 sets
- **3QFY06:** System 2 stabilization contract award
- **4QFY06:** CHALS-C contract award
- **4QFY06:** CHALS-C /GE/X-Midas location integration contract award
- **4QFY06:** X-Midas System 2 Audio contract award

Recent and Projected Activities

- **3QFY07:** ESA contract award
- **1QFY08:** GGB system fielding
- **2QFY08:** CHALS-C factory acceptance test
- **2QFY09:** ESA system assessment

ACQUISITION PHASE

Concept & Technology Development | System Development & Demonstration | Production & Deployment | Operations & Support

Guardrail Common Sensor (GR/CS)

FOREIGN MILITARY SALES
None

CONTRACTORS
Data links:
L-3 Communications (Salt Lake City, UT)
CHALS-C:
Lockheed Martin (Oswego, NY)
Guardian Eagle upgrades:
Northrop Grumman (Sacramento, CA)
Radio relay sets:
Raytheon (Falls Church, VA)
SIGINT support:
CACI (Eatontown, NJ)
Software/system support:
Northrop Grumman (Sacramento, CA)
X-Midas software:
ZETA (Fairfax, VA)

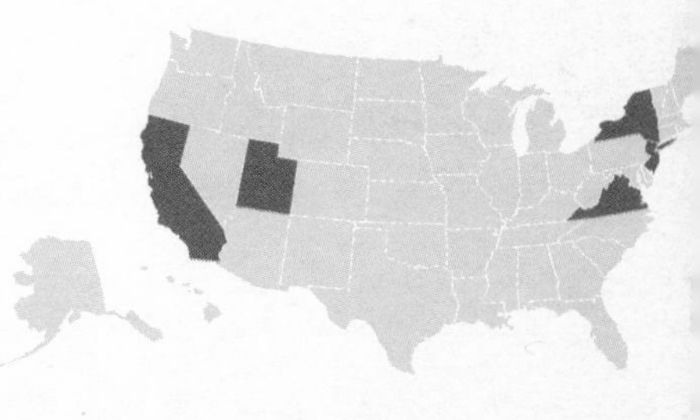

Guided Multiple Launch Rocket System (GMLRS)

Provides responsive, long-range, precision fires against area and point targets in open/complex/urban terrain with effects matched to the target and rules of engagement.

INVESTMENT COMPONENT

Modernization

Recapitalization

Maintenance

Description & Specifications

The Guided Multiple Launch Rocket System (GMLRS) is a major upgrade to the M26 rocket, producing precise destructive and shaping fires against a larger target set. GMLRS is employed with the M270A1 upgraded Multiple Launch Rocket System (MLRS) tracked launcher and the M142 High Mobility Artillery Rocket System (HIMARS) wheeled launchers.

GMLRS munitions have greater accuracy with a resulting higher probability of kill, smaller logistics footprint, and minimized collateral damage. There are two variants of the GMLRS: the dual-purpose improved conventional munitions (DPICM) variant (warhead consists of 404 small anti-personnel and anti-materiel grenades that are dispersed over the specific target); and the unitary variant (warhead consists of a single, 200-pound class high-explosive charge that provides blast and fragmentation effects on, above, or in a specific target). GMLRS DPICM development was an international cooperative program with the United Kingdom, Germany, France, and Italy. An urgent materiel release version of the GMLRS unitary variant has been produced and fielded in support of U.S. Central Command (CENTCOM) forces.

Rocket Length: 3,937mm
Rocket Diameter: 227mm
Rocket Reliability: Threshold 92 percent; objective: 95 percent
Ballistic Range(s): 15 to 70+ kilometers

Program Status

- **1–2QFY02:** Conducted successful early development test on GMLRS DPICM
- **4QFY02–1QFY03:** Conducted production qualification test on GMLRS DPICM
- **3QFY03:** Low rate initial production (LRIP) decision and LRIP I contract award for GMLRS DPICM
- **4QFY03:** FY03 Operation Iraqi Freedom (OIF) supplemental contract award
- **4QFY04:** Initial operational test of GMLRS DPICM
- **3QFY05:** Full-rate production decision for GMLRS DPICM
- **2QFY05–3QFY05:** Developmental testing on GMLRS Unitary UMR rocket
- **3QFY05:** Full rate production decision for GMLRS DPICM
- **3QFY05:** GMLRS Unitary UMR rocket fielded to CENTCOM forces
- **3QFY05–3QFY06:** Developmental testing conducted on GMLRS Unitary Objective rocket
- **2QFY06:** GMLRS DPICM receives type qualification
- **3QFY06:** Additional GMLRS Unitary UMR rockets fielded to OIF theater

Recent and Projected Activities

- **FY06–07:** Production qualification testing for GMLRS Unitary Objective rocket
- **2QFY07:** GMLRS Unitary Milestone C
- **FY08:** Initial operational test for GMLRS Unitary Objective rocket

ACQUISITION PHASE

Concept & Technology Development | System Development & Demonstration | Production & Deployment | Operations & Support

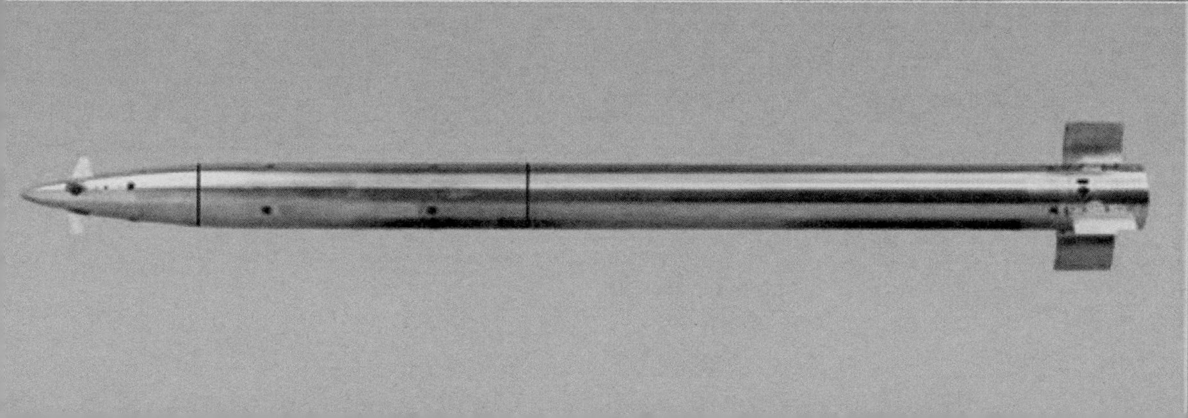

Guided Multiple Launch Rocket System (GMLRS)

FOREIGN MILITARY SALES
United Kingdom

CONTRACTORS
Prime munitions integrator:
Lockheed Martin (Dallas, TX)
Rocket assembly:
Lockheed Martin (Camden, AR)
Motor assembly:
Aerojet (Camden, AR)
G&C section:
Honeywell (Clearwater, FL)
Motor case/warhead skins:
Aerojet (Vernon, CA)

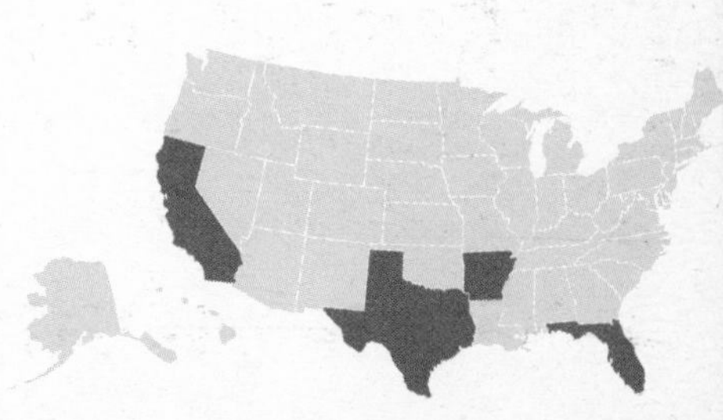

Heavy Expanded Mobility Tactical Truck (HEMTT) and HEMTT Extended Service Program (ESP)

Supports combat units by performing line and local haul, unit resupply, helicopter and tactical vehicle refueling, and related missions in a tactical environment.

INVESTMENT COMPONENT

- Modernization
- Recapitalization
- Maintenance

Description & Specifications

The rapidly deployable Heavy Expanded Mobility Tactical Truck (HEMTT), developed for cross-country military missions, transports ammunition, petroleum, oils, and lubricants to currently equipped, digitized, and brigade/battalion areas of operation. HEMTT is a prime mover for missile systems such as the PATRIOT and operates in any climate. A self-recovery winch is on all models.

The HEMTT A4 model includes air-ride suspension, a new C-15 engine, the Allison 4500 transmission, J-1939 data-bus, and a Palletized Load System (PLS)-size cab that will be common with the PLS A1 and Long Term Armor Strategy (LTAS)-compliant.

The HEMTT ESP is a recapitalization program that converts high-mileage HEMTT trucks to 0 Miles/0 Hours and to the current A2 and A4 (in FY08) production configurations. The trucks are disassembled and rebuilt with improved technology such as electronically controlled engine, electronic transmission, air ride seats, four-point seatbelts, bolt-together wheels, increased corrosion protection, enhanced electrical package, and air-ride suspension on the A4.

Program Status

- **Current:** Fielded to 82nd Airborne; 3rd Infantry Division (ID); 1-1nd Air Defense Artillery (ADA) Battalion (BN) (Patriot); 3-2 ADA BN (Patriot); several National Guard units
- **FY06:** HEMTT A4 development and testing; HEMTT A4 Production Verification Testing (PVT) at Aberdeen Test Center (ATC); HEMTT A4 vehicles used for LTAS
- **FY06:** HEMTT A3: Second prototype vehicle produced and tested at Oshkosh Truck Corp.; shakedown testing at Nevada Automotive Test Center (NATC); Platform System Demonstration (PSD) at ATC; C-130 direct off-load ability demonstrated at Oshkosh; Evaluation Vehicle (EV) in production

Recent and Projected Activities

- **FY07:** New contract award
- **FY07:** Projected fieldings: 75th Fires Brigade (BDE); 214th Fires BDE; 278th Armored Cavalry Regiment (ACR); 116th Armor BDE, ID; 256th AR BDE; 101st AASLT Aviation BDE; 10th Mountain Division, 4th ID, 2nd ID; National Guard units
- **FY07:** HEMTT A3. EV completed; performance testing (Oshkosh); shakedown test at ATC
- **FY07:** HEMTT A3. EV completed; performance testing (Oshkosh); shakedown test at ATC
- **3QFY07:** HEMTT A4 logistics demonstration and verification of technical manuals
- **4QFY07:** HEMTT A2 Light Equipment Transporter (LET) available
- **1QFY08:** HEMTT A4 Type classification/materiel release

ACQUISITION PHASE

- Concept & Technology Development
- System Development & Demonstration
- Production & Deployment
- Operations & Support

Heavy Expanded Mobility Tactical Truck (HEMTT) and HEMTT Extended Service Program (ESP)

FOREIGN MILITARY SALES
Turkey, Israel, and Jordan

CONTRACTORS
Oshkosh Truck Corp. (Oshkosh, WI)
Detroit Diesel (Emporia, KS; Redford, MI)
Allison Transmissions (Indianapolis, IN)
Michelin (Greenville, SC)

HEMTT comes in six basic configurations:

- M977 cargo truck with light materiel handling crane
- M985 cargo truck with materiel-handling is the ammunition transport prime mover for the Multiple-Launch Rocket System (MLRS)
- M978 is a 2,500-gallon fuel tanker
- M983 tractor
- M984 wrecker is a recovery vehicle for other vehicle systems
- M1120 load-handling system (LHS) transports palletized materiel and International Standards Organization (ISO) containers

Truck payload: 11 tons
Trailer payload: 11 tons
Flatrack dimensions: 8-foot-by-20-foot (ISO container standard)
Engine type: Diesel
Transmission: Automatic
Number of driven wheels: 8
Range: 300 miles
Air transportability: C-130, C-17, C-5

Hellfire Family of Missiles

Engages and defeats individual moving or stationary advanced-armor, mechanized or vehicular targets, patrol craft, buildings or bunkers while increasing aircraft survivability.

INVESTMENT COMPONENT

Modernization

Recapitalization

Maintenance

Description & Specifications

The Hellfire family of munitions, consisting of the AGM-114 A, C, F, K, L, M and N model missiles, provides air-to-ground precision strikes and is designed to defeat individual hard point targets. The Laser Hellfire (Hellfire II) comes with either a shaped charge warhead for defeating armor targets or a penetrating blast fragmentation warhead for defeating buildings and bunkers. It uses semi-active laser terminal guidance and is the primary anti-tank armament of the AH-64 Apache, OH-58 Kiowa Warrior, Armed Recon Helicopter, special operations aircraft, the U.S. Marine Corps' AH-1W Super Cobra Helicopters, and the Army's Warrior Unmanned Aerial System (UAS).

The Longbow Hellfire (L model–no longer in production) uses millimeter wave technology for terminal guidance. The Hellfire's ability to engage single or multiple targets directly or indirectly and to fire single, rapid, or ripple (salvo) rounds gives combined arms forces a decisive battlefield advantage.

Longbow Hellfire:
Diameter: 7 inches; Weight: 108 pounds; Length: 69.2 inches; Range: 0.50 – 8.0 kilometers

Laser Hellfire (AGM-114K, M, and N models) and Longbow Hellfire incoporate many improvements over the basic Hellfire missile, including:

- Electro Optic Countermeasure Hardening
- Software-controlled digital seeker and autopilot electronics to adapt to changing threats and mission requirements
- Increased warhead lethality capable of defeating all projected armor threats into the 21st century

Laser Hellfire's semi-active laser precision guidance and Longbow Hellfire's fire-and-forget capability will provide the battlefield commander with fast battlefield response and flexibility across a wide range of mission scenarios.

Laser Hellfire
Diameter: 7 inches; Weight: 100 pounds; Length: 64 inches; Range: 0.50–8.0 kilometers

Program Status

Laser Hellfire

- **4QFY04–3QFY06:** Completed FY03 production delivery
- **1QFY06:** Production deliveries; transition to production of AGM-114 variants
- **2QFY06–3QFY06:** Completed FY04 production delivery

Longbow Hellfire

- **3QFY04–4QFY06:** Completed final multi-year deliveries

Recent and Projected Activities

Longbow Hellfire

- Continue sustainment activities

ACQUISITION PHASE

Concept & Technology Development | System Development & Demonstration | **Production & Deployment** | Operations & Support

Production Awards	Characteristics	Performance	System Description
1990-1992	• Dual Warhead • SAL Seeker • Analog Autopilot	• Reactive Armor • Non-EOCM Hardened • Not Programmable for Improved Performance	*Interim* (IHW) (AGM-114F) Weight = 107 lbs Length = 71"
1993-1997	• Robust Warheads • EOCM Hardened • SAL Seeker • Reprogrammable • 114K2 IM Warheads - Apr 97	• Capable Against Predicted Early 21st Century Armor • 114K2A add Blast Frag sleeve	*HELLFIRE II* (AGM-114K, K2, Weight = 100 lbs Length = 64"
1996-	• Robust Warheads • RF Seeker • Reprogrammable	• Capable Against Predicted Early 21st Century Armor • All Weather • Fire and Forget	*LONGBOW* (AGM-114L) Weight = 108 lbs Length = 69"
1999-	• IM Warhead • SAL Seeker • Reprogrammable	• Steel/Masonry • Digital autopilot and electronics • Blast frag warhead with electronic S/A delay fuse	*Blast Frag* (AGM-114M) Weight = 106 lbs Length = 64"
2003	• MAC Warhead (Metal Augmented Charge) • No Precursor SC • Main Penetrating Blast Frag	• Delay Fuzing • SAL Guidance • Urban Structures, Bunkers, Caves, Personnel	*HELLFIRE II MAC* (AGM-114N) Weight = 106 lbs Length = 64"
2008	• Robust Warheads • EOCM Hardened • SAL Seeker • Reprogrammable	• Enhanced Software for UAV Engagements • Multiple warheads available • Backward Compatible on Current Force Platforms	*HELLFIRE II High Altitude(AGM-114)* Weight = 99 lbs Length = 64" Weight = 104 lbs w/ Steel Frag Sleeve Weight = 106 lbs (MAC/BF Warhead)

Hellfire Family of Missiles

FOREIGN MILITARY SALES

Laser Hellfire:
Singapore, Israel, Kuwait, Netherlands, Greece, Egypt, Saudi Arabia, Taiwan, Australia, and Spain

Direct commercial sale: United Kingdom

Longbow Hellfire:
Singapore, Israel, Kuwait, and Japan

Direct commercial sale: United Kingdom

CONTRACTORS

Laser Hellfire Missile System, guidance section, sensor group:
HELLFIRE LLC (Orlando, FL)

Longbow Hellfire:
Longbow LLC (Orlando, FL)

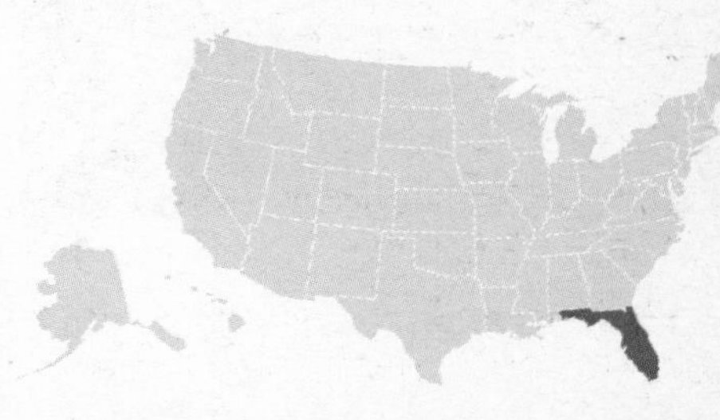

High Mobility Artillery Rocket System (HIMARS)

Enables early entry and contingency forces to engage and defeat area and point targets in both urban/complex and open terrain at long ranges, via a highly mobile and deployable multiple launch, precision rocket and missile system.

INVESTMENT COMPONENT

- Modernization
- Recapitalization
- Maintenance

Description & Specifications

The combat-proven High Mobility Artillery Rocket System (HIMARS) provides the Army and Marine Corps with a rapidly deployable—C-130 transportable—round-the-clock, all-weather, lethal, long-range precision rocket and missile fire support system for joint, early entry expeditionary forces, contingency forces, and modular fire brigades in support of Brigade Combat Teams.

Highly mobile, HIMARS is mounted on a five-ton modified Family of Medium Tactical Vehicles chassis. The wheeled chassis allows for faster road movement, and lower operating costs, and requires 30 percent fewer strategic airlifts (via C-5 or C-17) to transport than the current, tracked M270 Multiple Launch Rocket System that it replaces. HIMARS can fire all current and planned suites of Multiple Launch Rocket System family of munitions, including Army Tactical Missile System missiles and Guided Multiple Launch Rocket System rockets. HIMARS carries either six rockets or one missile, is self-loading and self-locating, and is operated by a three-man crew protected during firings in either a reinforced man-rated cab or an armored cab.

Program Status

- **2QFY05:** First unit equipped, 3rd Battalion, 27th Field Artillery Regiment, 18th Fires Brigade
- **3QFY05:** Full rate production decision review
- **1QFY06:** Full rate production I contract awarded
- **2QFY06:** Second unit equipped, 1st Battalion, 181st Field Artillery Regiment, 138th Fires Brigade
- **1QFY07:** Full rate production II contract award

Recent and Projected Activities

- Continue fielding to active and reserve components
- Provide support to fielded units

ACQUISITION PHASE

Concept & Technology Development | System Development & Demonstration | Production & Deployment | Operations & Support

High Mobility Artillery Rocket System (HIMARS)

FOREIGN MILITARY SALES
None

CONTRACTORS

Prime and launcher:
Lockheed Martin (Dallas, TX; Camden, AR)

Family of Medium Tactical Vehicles:
Armor Holdings Aerospace & Defense Group (Sealy, TX)

Improved Weapons Interface Unit:
Harris Corp. (Melbourne, FL)

Position Navigation Unit:
L-3 Communications Space & Navigation (Budd Lake, NJ)

Hydraulic pump and motor:
Vickers (Jackson, MS)

High Mobility Multipurpose Wheeled Vehicle (HMMWV)

Supports combat, combat support, and combat service support units with a versatile, light, mission-configurable, tactical wheeled vehicle.

INVESTMENT COMPONENT

- Modernization
- **Recapitalization**
- Maintenance

Description & Specifications

The High Mobility Multipurpose Wheeled Vehicle (HMMWV) is a tri-service program that provides light, highly mobile, diesel-powered, four-wheel-drive vehicles to satisfy Army, Marine Corps, and Air Force requirements. The HMMWV uses common components to enable its reconfiguration as a troop carrier, armament carrier, shelter carrier, ambulance, TOW missile carrier, and scout vehicle. Since its inception, the HMMWV has undergone numerous improvements, including technological upgrades, higher payload capacity, radial tires, Environmental Protection Agency emissions update, commercial bucket seats, three-point seat belts and other safety enhancements, four-speed transmissions, and in some cases, turbocharged engines and air conditioning.

There are numerous HMMWV variants. The HMMWV A2 configuration incorporates the four-speed, electronic transmission, the 6.5-liter diesel engine, and improvements in transportability. It serves as a platform for other Army systems such as the Ground-Based Common Sensor. The heavy variant has a payload of 4,400 pounds and is the prime mover for the light howitzer and heavier shelters. The expanded capacity vehicle (ECV) has a payload capacity of 5,100 pounds, including crew and kits. The ECV chassis serves as a platform for mission payloads and for systems that exceed 4,400 pounds and is used for the M1114 Up-armored HMMWV. The Up-Armored HMMWV was developed to provide increased ballistic and blast protection, primarily for military police, special operations, and contingency force use.

The M1151 Armament Carrier, the M1152 (2-door variant) Troop/Cargo/Shelter Carrier, and the M1165 (4-door variant) Command and Control Carrier have all been recently introduced into production. These variants are also built on an ECV chassis, which provides additional carrying capacity and are produced with an integrated armor package (A- Kit) with the capability to accept add-on-armor kits (B-Kits). The M1151 is currently fielded with a gunner's protection kit with transparent armor.

The HMMWV recapitalization program creates vehicles with increased capability, reliability, and maintainability.

Program Status

- Continue fielding of ECV HMMWVs to Army, Marine Corps, Air Force, and foreign military sales customers
- Continue recapitalization of HMMWVs
- **1QFY07:** Full materiel release for M1151A1/1152A1/1165A1 expected

Recent and Projected Activities

- Continuous product improvements through the introduction of upgraded components

ACQUISITION PHASE

- Concept & Technology Development
- System Development & Demonstration
- **Production & Deployment**
- Operations & Support

High Mobility Multipurpose Wheeled Vehicle (HMMWV)

FOREIGN MILITARY SALES
Argentina, Bahrain, Bolivia, Chad, Colombia, Djibouti, Ecuador, Egypt, Ethiopia, Honduras, Israel, Kuwait, Luxembourg, Mexico, Oman, Philippines, Saudi Arabia, Sudan, Taiwan, Tanzania, Tunisia, and Uganda

CONTRACTORS
AM General (South Bend, IN)
Armor Holdings Inc. (AHI) (Fairfield, OH)
GEP (Franklin, OH)
Defiance (Defiance, OH)
GM (Warren, MI)
Red River Army Depot (Red River, TX)
Letterkenny Army Depot (Chambersburg, PA)
Maine Military Authority (Limestone, ME)

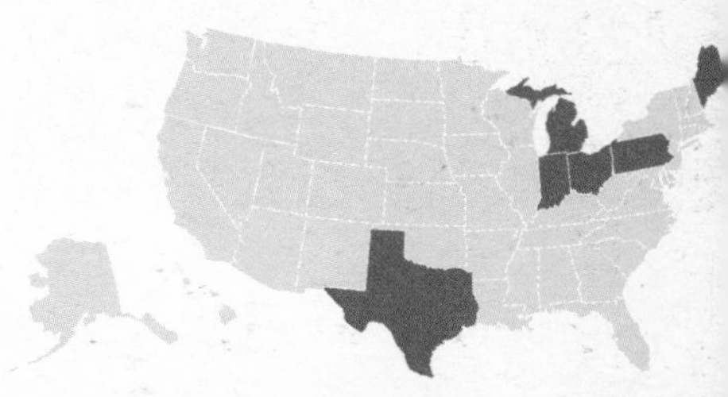

Improved Ribbon Bridge (IRB)

Improves mobility by providing continuous roadway or raft capable of crossing military load classification 96 (wheeled)/80 (tracked) vehicles over non-fordable wet gaps.

INVESTMENT COMPONENT

Modernization

Recapitalization

Maintenance

Description & Specifications

The Tactical Float Ribbon Bridge System consists of the Improved Ribbon Bridge (IRB) bays (30 interior and 12 ramps); 14 Propulsion Bridge Erection Boats (BEB), and 56 Common Bridge Transporters (CBT). These components are required to transport, launch, erect, and retrieve up to 210 meters of floating bridge. The IRB has a Military Load Capacity (MLC) of 96 wheeled (normal) and 110 (caution)/MLC 80 tracked vehicles and is used to transport weapon systems, troops, and supplies over water when permanent bridges are not available. This MLC will support the joint force commander's ability to employ and sustain forces throughout the world. The Float Ribbon Bridge is issued to the Multi-Role Bridge Company (MRBC).

The M14 Improved Boat Cradle (IBC) and the M15 Bridge Adapter Pallet (BAP) are used to carry the BEB and IRB bays on the CBT.

Program Status

- **Current:** This system has been fielded since 2004

Recent and Projected Activities

- Fieldings are ongoing based on the Army Requirements Prioritization List (ARPL)

ACQUISITION PHASE

Concept & Technology Development | System Development & Demonstration | Production & Deployment | Operations & Support

Improved Ribbon Bridge (IRB)

FOREIGN MILITARY SALES
None

CONTRACTORS
IRB Bays manufacturer:
General Dynamics Santa Barbara Sistemas (Kaiserslautern, Germany)
Logistic support:
AM General (Livonia, MI)
CBT manufacturer:
Oshkosh Truck Corp. (Oshkosh, WI)
BEB manufacturer:
FBM Babcock Marine (Isle of Wight, United Kingdom)

Improved Target Acquisition System (ITAS)

Provides superior surveillance capability that enables the Soldier to shape the battlefield by detecting and engaging targets at long ranges.

INVESTMENT COMPONENT

Modernization

Recapitalization

Maintenance

Description & Specifications

The TOW Improved Target Acquisition System (ITAS) is a reconnaissance, surveillance, and target acquisition (RSTA) platform, providing long-range heavy anti-tank and precision assault fires capabilities to Army and Marine Corps light, airborne, air assault, and Stryker Brigade Combat Team (SBCT) forces. It is a multi-mission weapon system used as a tank neutralizer, precision assault weapon, and as the infantry task force's long-range surveillance asset.

ITAS is a major product upgrade that greatly reduces its number of components, minimizing logistics support and equipment requirements. Built-in diagnostics and improved interfaces enhance target engagement performance. ITAS's second-generation infrared sensors double the range of its predecessor for gunners. It offers improved hit probability by aided target tracking, improved missile flight software algorithms, and an elevation brake to minimize launch transients.

The ITAS includes an integrated (day/night sight with laser rangefinder) target acquisition subsystem, a fire control subsystem, a lithium-ion battery power source, and a modified traversing unit. Soldiers can detect and engage long-range targets with TOW missiles or direct fire to destroy them. With the PAQ-4/PEQ-2 Laser Pointer, ITAS can designate .50 caliber or Mk-19 grenade engagements. TOW 2B Aero provides an extended maximum range to 4500 meters for long-range engagement of armored vehicles. The TOW Bunker Buster is used in urban terrain or bunker engagements in assault operations. ITAS operates from the HMMWV, the dismount tripod platform, and Stryker anti-tank guided missile vehicles.

Program Status

- **Current:** ITAS has been fielded to 82d Airborne Div; 1st, 2nd and 3rd SBCTs; the 101st Air Assault Division; 2nd Infantry, 10th Mountain, and 25th Infantry Divisions; 39th, Separate Infantry Brigade; 1/34th Minnesota National Guard.
- **Current:** The Marine Corps selected ITAS to replace all Marine Corps heavy anti-tank/assault systems and placed 418 systems on contract on September 12, 2006.
- **2QFY05–4QFY06:** Fielding of Army pre-positioned stock; fielding of SBCT 1 and SBCT 4 in-lieu-ofs; modularity conversion of 11 BCTs; retrofitted 205 silver-zinc batteries with lithium-ion batteries; reset of units returning from Theater of Operation; sustainment training for fielded units; pre-deployment training; National Guard support of border Patrol activities; contractor logistics support

Recent and Projected Activities

- **2QFY07–2QFY09:** Fielding of 11 BCTs; completion of lithium-ion retrofit; ITAS production continues; sustainment training for fielded units; pre-deployment training; anticipated continuation of border patrol activities; contractor logistics support

ACQUISITION PHASE

Concept & Technology Development	System Development & Demonstration	Production & Deployment	Operations & Support

Improved Target Acquisition System (ITAS)

FOREIGN MILITARY SALES
NATO Maintenance and Supply Agency
Canada

CONTRACTORS
Prime:
Raytheon (McKinney, TX)
Training Devices:
Intercoastal Electronics (Mesa, AZ)

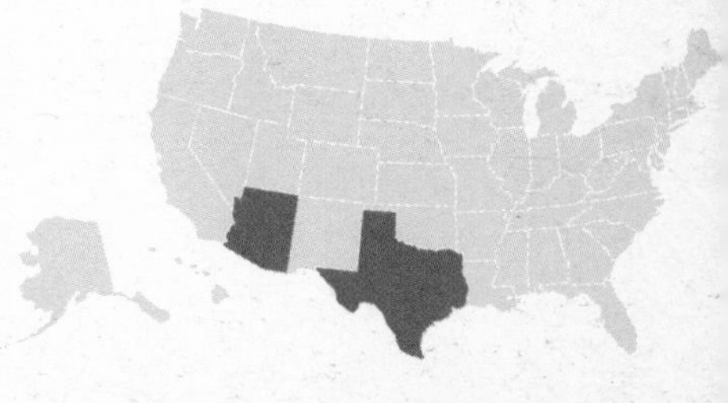

Installation Protection Program (IPP) Family of Systems

Provides an effective chemical, biological, radiological, and nuclear (CBRN) protection, detection, identification, and warning system for military installations.

INVESTMENT COMPONENT

Modernization

Recapitalization

Maintenance

Description & Specifications

The Installation Protection Program (IPP) will allow DoD military installations to effectively protect critical operations and personnel against a chemical, biological, radiological, and nuclear (CBRN) event, to effectively respond with trained and equipped emergency responders, and to ensure installations can continue critical operations during and after an attack.

The IPP uses a tiered approach of government and commercial off-the-shelf capabilities optimized for an installation. The Tiers are as follows:

- Baseline Tier provides a set of non-materiel support tools tailored for each Service that consist of training products, planning templates, Mutual Aid Agreement (MAA) templates, and exercise templates and scenarios. It also includes funding for the development, testing and improvement of DoD CBRN installation protection doctrine, policy and standards, equipment, and systems. This includes the development of Computer Based Training (CBT) for civilian awareness, exercise templates and scenarios and memorandum of understanding (MOU) and MAA templates, which focus specifically on CBRN-related response requirements.
- Tier 1 provides the Baseline Tier capabilities as well as individual protective equipment (IPE) for emergency responders and first receivers; portable radiological, chemical and biological detection equipment; personal dosimeters; hazard marking and controlling equipment; medical countermeasures for emergency responders, first receivers and first responders; mass casualty decontamination showers and tents; mass casualty litters and support equipment; mass notification systems; decision support tools to include hand-held computers; incident management software; new equipment training; and table-top and field exercise support.
- Tier 2 provides the Baseline Tier, Tier 1, and fixed chemical detectors for warfare agents and toxic industrial materials and chemicals; fixed biological collectors with analysis and identification laboratory support; collective protection for one of a kind strategic assets (up to 10,000 square feet); escape masks to permit evacuation of personnel identified to work in collective protection; and a decision support system.

Program Status

- **1QFY03:** Program initiated
- **1QFY06:** Funding reduced; changed program strategy
- **2QFY06:** Begin fielding new (IPP Lite) program strategy
- **1QFY07:** Field 36 installations

Recent and Projected Activities

- **1QFY08:** Complete nine OCONUS installations

ACQUISITION PHASE

Concept & Technology Development | System Development & Demonstration | Production & Deployment | Operations & Support

Installation Protection Program (IPP) Family of Systems

FOREIGN MILITARY SALES
None

CONTRACTORS
Lead systems integrator:
SAIC (Falls Church, VA)

Integrated Family of Test Equipment (IFTE)

Enables verification of Army weapon systems and components, and isolates/diagnoses and repairs faults through mobile, general purpose, and automatic test systems.

INVESTMENT COMPONENT

Modernization

Recapitalization

Maintenance

Description & Specifications

The Integrated Family of Test Equipment (IFTE) consists of two interrelated, integrated, mobile, tactical, and man-portable systems. These rugged, compact, lightweight, general purpose systems enable verification of the operational status of weapon systems, as well as fault isolation to the line replaceable unit at all maintenance levels, both on and off the weapon system platform.

Base Shop Test Facilities (BSTF): The Electro-Optics Test Facility (EOTF) (AN/TSM-191(V)5) tests the full range of Army electro-optical systems including laser transmitters, receivers, spot trackers, forward-looking infrared systems, and television systems. It is fully mobile with VXI instrumentation, touch-screen operator interface, and an optical disk system for test program software and electronic technical manuals.

Next Generation Automatic Test System (NGATS) facility (AN/TSM-191(V)6) is the follow-on reconfigurable, rapidly deployable, automatic test equipment that supports joint operations, reduces logistical footprint, and replaces/consolidates obsolete, unsupportable automatic test equipment in the Army inventory.

Maintenance Support Device–Version 2 (MSD-V2): MSD-V2, the second generation MSD, is a lightweight, rugged, compact, man-portable general-purpose automatic tester used to verify the operational status of weapon systems, both electronic and automotive, and to isolate faulty components for immediate repair and/or replacement. MSD-V2 hosts Interactive Electronic Technical Manuals, is used as a software uploader/verifier to provide or restore mission software to weapon systems, and supports testing and diagnostic requirements of current and Future Combat Systems (FCS). MSD-V2 supports more than 40 weapon systems and is used by more than 25 maintainer military occupational specialties.

Program Status

- **1QFY06:** MSD-V2 Fielding ongoing (production and deployment)
- **1QFY06:** NGATS system development and demonstration (Increment 1)
- **1QFY07:** NGATS system development and demonstration (Increment 2)

Recent and Projected Activities

- **2QFY07:** EOTF production and deployment continues
- **2QFY07:** NGATS developmental testing
- **1QFY08:** MSD-Version 3 (MSD-V3) contract award
- **3QFY08:** NGATS operational testing
- **4QFY09:** NGATS first unit equipped

ACQUISITION PHASE

Concept & Technology Development | System Development & Demonstration | Production & Deployment | Operations & Support

AN/TSM-191(V) 5 Electro-Optics Test Facility

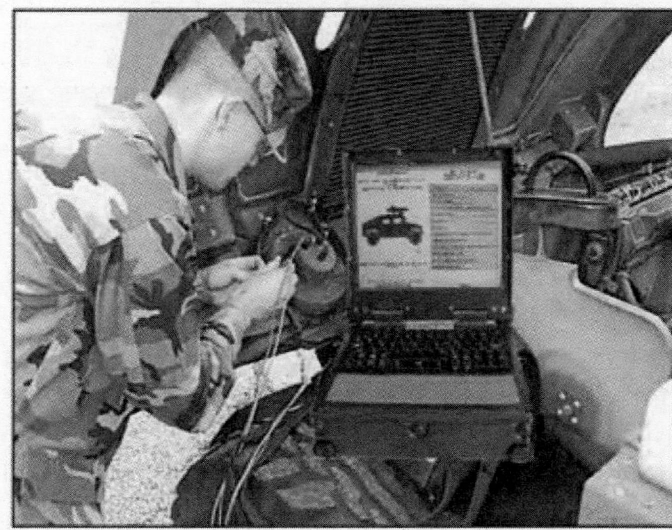

Maintenance Support Device

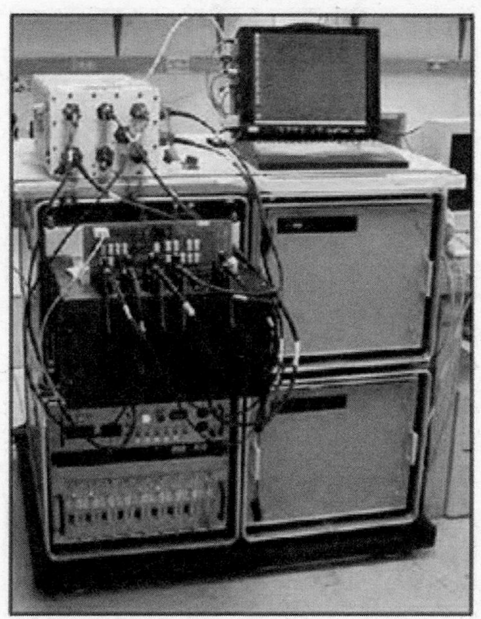

AN/TSM-191(V) 6 Base Shop Test Facility

Integrated Family of Test Equipment (IFTE)

FOREIGN MILITARY SALES

Australia, Afghanistan, Bahrain, Chile, Djibouti, Egypt, Ethiopia, Germany, Israel, Iraq, Jordan, Korea, Kuwait, Lithuania, Macedonia, Morocco, Netherland, Oman, Poland, Portugal, Saudi Arabia, Taiwan, Turkey, United Arab Emirates, Uzbekistan, Yemen

CONTRACTORS

EOTF:
Northrop Grumman (Rolling Meadows, IL)
MSD-V2:
Science & Engineering Services, Inc. (SESI) (Huntsville, AL)
Vision Technology Miltope Corp. (Hope Hull, AL)
BSTF:
DRS Technologies (Huntsville, AL)
Northrop Grumman (Rolling Meadows, IL)

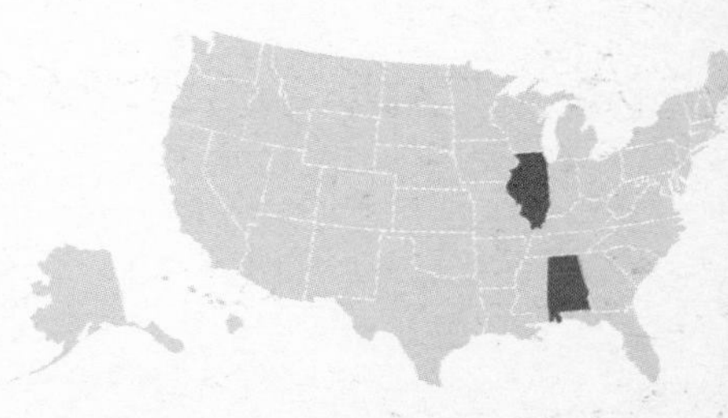

Integrated System Control (ISYSCON) (V)4/Tactical Internet Management System (TIMS)

Improves command and control, planning, and engineering of the Army's tactical networks, from battalion through theater, in support of joint and combined operations.

INVESTMENT COMPONENT

Modernization

Recapitalization

Maintenance

Description & Specifications

The Integrated System Control (ISYSCON) (V)4/Tactical Internet Management System (TIMS) is a software system that resides on the Force XXI Battle Command Brigade-and-Below (FBCB2) system located in the S6/G6 sections of the digitized force architecture. The ISYSCON (V)4/TIMS reuses FBCB2 software as a foundation and adds developmental and commercial, off-the-shelf software to plan, configure, initialize, and monitor the Tactical Internet. The ISYSCON (V)4/TIMS enhances the FBCB2 system management capability.

The ISYSCON (V)4/TIMS is a command and control enabler that will support the full spectrum of military operations and the seven mission areas described in the Army Planning Guidance sections of The Army Plan FY02–10. The ISYSCON (V)4/TIMS is intended to be developed and implemented in increments by incorporating blocked enhancements to the key performance parameter threshold baseline.

The ISYSCON (V)4/TIMS is expected to evolve into new hardware and software baseline blocked enhancements. As new systems are added to the Tactical Internet, such as aviation platforms, the ISYSCON (V)4/TIMS will provide a network management interface capability. The ISYSCON (V)4/TIMS is scheduled for fielding to the active duty components and 15 enhanced separate brigades.

Program Status

- **3QFY05:** Commence development of initialization capability
- **1–4QFY05:** Field in accordance with Unit Set Fielding Schedule (includes elements of 4th Infantry Division (ID), 10th Mountain Division, 101st Airborne (ABN), and Stryker Brigade Combat Team (SBCT) 5)
- **1–4QFY06:** Field in accordance with Unit Set Fielding Schedule (includes elements of 1/34th, 1st Cavalry Division, 82nd ABN, 1st ID, 2nd ID, 3rd ID, 25th ID, SBCT 5 and SBCT 6)
- **4QFY06:** Developmental testing for Software Block 2 software baseline
- **1QFY07:** Commenced development of Joint Capabilities Release (JCR) software baseline

Recent and Projected Activities

- **1Q–4QFY07:** Continue fielding in accordance with Unit Set Fielding Schedule
- **3QFY07:** Intra Army interoperability certification testing for Software Block 2 baseline
- **2QFY08:** Commence fielding of initialization capability
- **3QFY08:** Intra Army interoperability certification testing for Software Block 2+ baseline
- **1Q–4QFY08:** Prepare for and transition of software for post-production software support
- **1QFY09:** Continue to field initialization capability

ACQUISITION PHASE

Concept & Technology Development | System Development & Demonstration | Production & Deployment | Operations & Support

Integrated System Control (ISYSCON) (V)4/Tactical Internet Management System (TIMS)

FOREIGN MILITARY SALES
None

CONTRACTORS
Software:
Northrop Grumman Mission Systems (Carson, CA)
CSC (Falls Church, VA)
Hardware:
GTSI (Chantilly, VA)
Integrated logistics support, fielding & training:
Engineering Solutions & Products (Eatontown, NJ)

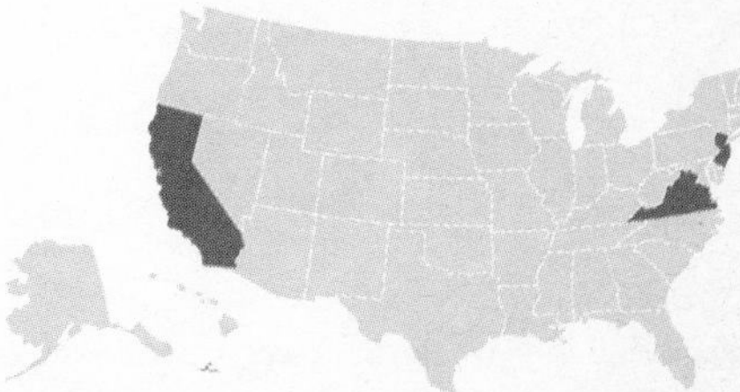

Interceptor Body Armor (IBA)

Protects individual Soldiers from ballistic and fragmentation threats in a lightweight, modular body armor package.

INVESTMENT COMPONENT

Modernization

Recapitalization

Maintenance

Description & Specifications

Interceptor Body Armor (IBA) is modular, multiple-threat body armor, consisting of:

- Outer Tactical Vest (OTV)
- Enhanced Small Arms Protective Inserts (ESAPI)
- Enhanced Side Ballistic Inserts (ESBI)
- Deltoid and Axillary Protector (DAP)

Eight sizes of OTVs and five sizes of ESAPI plates are being fielded. The basic system weight (OTV and SAPI, size: medium) is 16.7 pounds, or more than 9 pounds lighter than the Interim Small Arms Protective Overvest (ISAPO) and the Personnel Armor System Ground Troops (PASGT) combination, which it replaces. The medium size OTV, without plates, weighs 7.7 pounds and protects against fragmentation and 9-millimeter rounds. The ESAPI plates provide additional protection and can withstand multiple small arms hits. IBA includes attachable throat, groin, and collar protectors for increased protection, and webbing attachment loops on the front and back of the vest for attaching pouches for the Modular Lightweight Load-Carrying Equipment (MOLLE). DAP provides additional protection from fragmentary and 9-millimeter projectiles to the upper arm and underarm areas. During Operation Iraqi Freedom combat operations, the side and underarm areas not currently covered by the Enhanced Small Arms Protective Insert (ESAPI) component of the IBA were identified by combat commanders and medical personnel as a vulnerability that needed to be addressed. To meet this threat and provide an increased level of protection, the ESBI was developed. These features allow commanders the flexibility to tailor the IBA to meet the specific mission needs or changing threat conditions.

Program Status

- **Current:** In production and being fielded
- **4QFY06:** 964,000 OTVs and 964,000 ballistic insert sets fielded

Recent and Projected Activities

- Continue fielding

ACQUISITION PHASE

Concept & Technology Development | System Development & Demonstration | Production & Deployment | Operations & Support

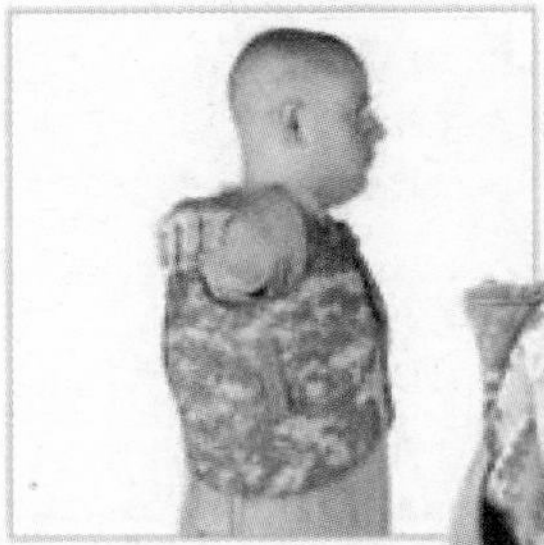

Interceptor Body Armor (IBA)

FOREIGN MILITARY SALES
None

CONTRACTORS
Point Blank Body Armor (Pompano Beach, FL)
Armacel Armor (Camarillo, CA)
ArmorWorks (Tempe, AZ)
Ceradyne, Inc. (Costa Mesa, CA)
Armor Holdings/Simula Safety Systems, Inc. (Phoenix, AZ)
Armor Holdings/Specialty Defense Systems (Alexandria, VA)
UNICOR
Protective Materials Company (Miami Lakes, FL)

Javelin

Improves combat effectiveness, lethality, and survivability of infantry, scouts, and combat engineers via a man-portable, fire-and-forget, medium anti-tank weapon.

INVESTMENT COMPONENT

Modernization

Recapitalization

Maintenance

Description & Specifications

Javelin is the first fire-and-forget, shoulder-fired, anti-tank missile fielded to the Army and Marine Corps, replacing Dragon. Javelin's unique top-attack flight mode, self-guiding tracking system, and advanced warhead design enables it to defeat all known tanks out to ranges of 2,500 meters.

Javelin's two major modular components are a reusable command launch unit and a missile sealed in a disposable launch tube assembly. The command launch unit's integrated day/night sight provides target engagement capability in adverse weather and countermeasure environments. The command launch unit also may be used by itself for battlefield surveillance and reconnaissance.

The Javelin missile and command launch unit together weigh 49.5 pounds. Its fire-and-forget capability enables gunners to fire and then immediately take cover, greatly increasing survivability. Special features include a selectable top-attack or direct-fire mode (for targets under cover or for use in urban terrain against bunkers and buildings), target lock-on before launch, and a very limited back-blast that enables gunners to fire safely from enclosures and covered fighting positions. Its maximum range is in excess of 2,500 meters.

The Javelin Block I program provides lethal performance essential to the current and objective forces by aligning the command launch unit and missile performance for maximum effectiveness at extended ranges. The Javelin weapon system is the dismounted weapon of choice for the Future Combat Systems (FCS) Unit of Action complementary systems. Javelin Block I was made available for FCS integration in FY07.

Javelin has been fielded to more than 90 percent of active duty units. Fielding is under way to the National Guard.

Program Status

- **FY04:** U.S. Marine Corps fully fielded
- **FY05:** Hardware fieldings to the Army continue on schedule
- **FY06:** Hardware contract in place to provide Block I
- **1QFY07:** 1st Brigade Combat Team (BCT) 2nd Infantry (ID) Division, BCT conversion; Retraining of 116th BCT ID Army National Guard (ARNG), 256th BCT LA ARNG, and 278 BCT TN ARNG

Recent and Projected Activities

- **FY07–09:** Procurement
- **2QFY07:** 156th BCT conversion; MS ARNG 82nd BCT conversion
- **3QFY07:** Conversion of BCT VA ARNG, 27th BCT NY ARNG, 1st ID BCT
- **4QFY07:** Retraining of 1-509th Infantry Joint Readiness Training Center, 11th Armor Cavalry, Ft. Irwin, CA

ACQUISITION PHASE

Concept & Technology Development | System Development & Demonstration | **Production & Deployment** | Operations & Support

Javelin

FOREIGN MILITARY SALES
United Kingdom, Australia, Ireland, Jordan, Lithuania, Taiwan, Norway, New Zealand, Czech Republic, and Oman

CONTRACTORS
Javelin joint venture:
Raytheon (Tucson, AZ) and Lockheed Martin (Orlando, FL)

Joint Biological Agent Identification Diagnostic System (JBAIDS)

Protects personnel by rapidly identifying low levels of biological warfare agents and other pathogens encountered in field or hospital clinical laboratories.

INVESTMENT COMPONENT

Modernization

Recapitalization

Maintenance

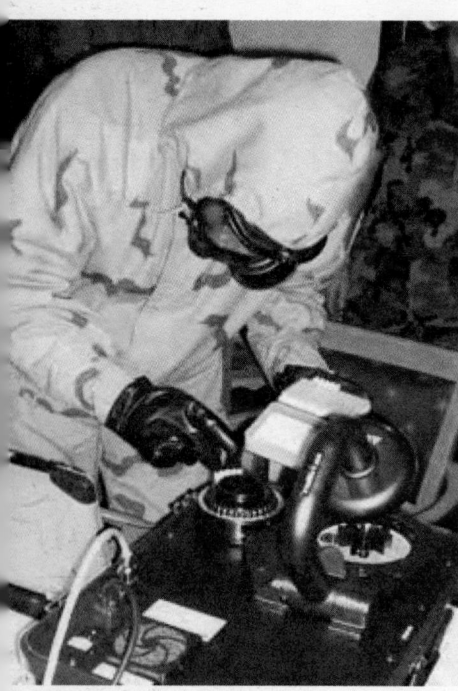

Description & Specifications

The Joint Biological Agent Identification Diagnostic System (JBAIDS) can identify biological agents in a variety of environmental and clinical samples at or below 1,000 colony-forming units or 10,000 plaque-forming units per milliliter.

JBAIDS includes a clinical instrument based on commercial, off-the-shelf, non-developmental item technology requiring limited modification to meet operational requirements. Integrated with this instrument are reagent test kits for pathogen identification and protocols for sample preparation that, when used together, can be cleared by the Food and Drug Administration as a diagnostic test.

- Detection Sensitivity: Equal to or better than 85 percent for identification of target agents at specified limit of detection concentrations
- Detection Specificity: Equal to or better than 90 percent for identification of target agents at specified limit of detection concentrations
- JBAIDS Block II upgrade will add the capability to identify toxins

Program Status

- **2QFY04:** Critical design review
- **2QFY04:** Developmental testing
- **4QFY04:** Operational assessment; initial production decision
- **4QFY05:** JBAIDS Block II competitive fly-off
- **2QFY06:** Full rate production decision (Ground Systems)

Recent and Projected Activities

- **3QFY07:** Follow-on test and evaluation
- **4QFY07:** Full rate production decision (Naval Systems)

ACQUISITION PHASE

Concept & Technology Development | System Development & Demonstration | Production & Deployment | Operations & Support

Joint Biological Agent Identification Diagnostic System (JBAIDS)

FOREIGN MILITARY SALES
None

CONTRACTORS
Idaho Technologies (Salt Lake City, UT)

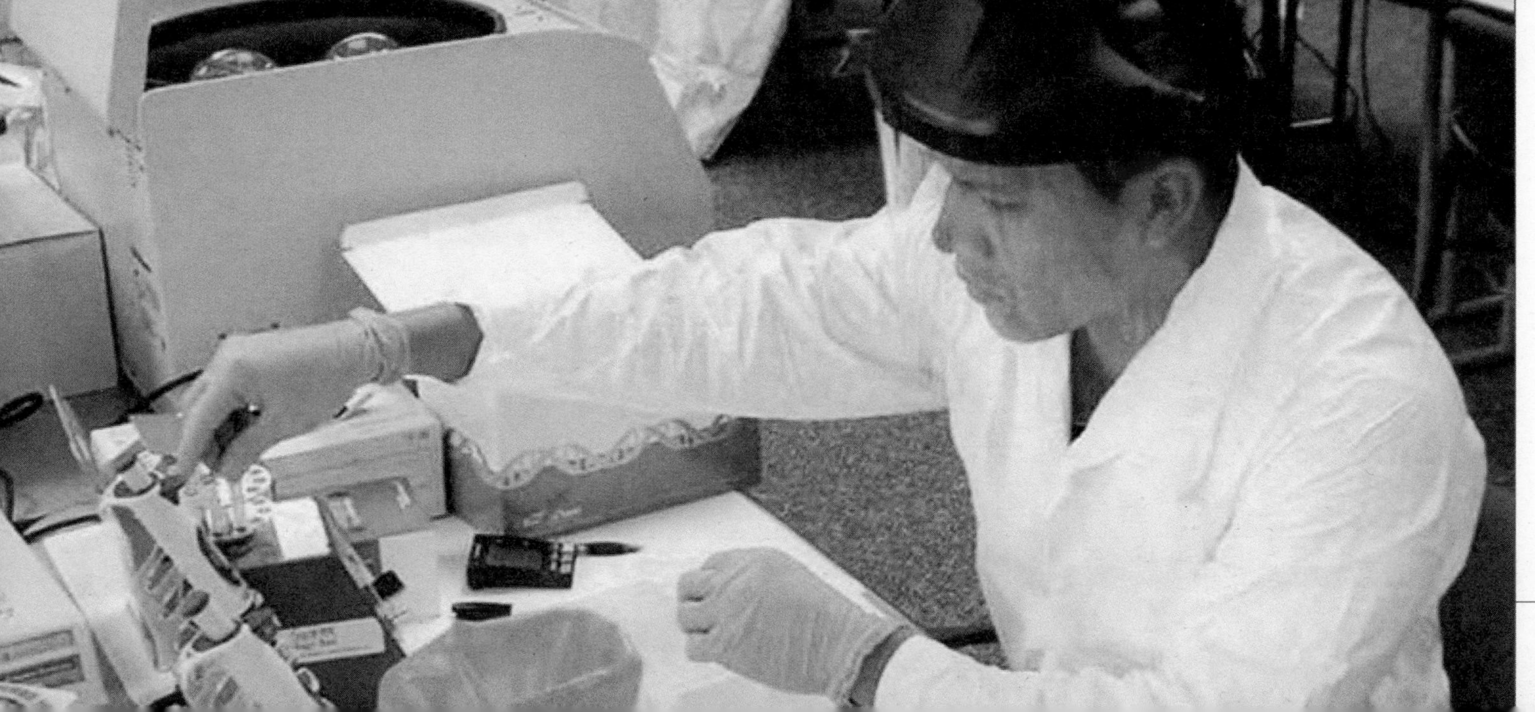

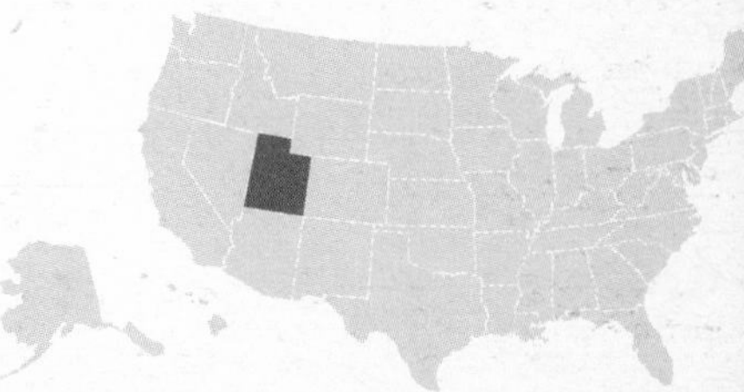

Joint Biological Point Detection System (JBPDS)

Protects the Soldier by providing rapid and fully automated detection, identification, warning, and sample isolation of high-threat biological warfare agents.

INVESTMENT COMPONENT

- **Modernization**
- Recapitalization
- Maintenance

Description & Specifications

The Joint Biological Point Detection System (JBPDS) is the first joint biological warfare (BW) agent detection system designed to meet the broad spectrum of operational requirements encountered by the services, across the entire spectrum of conflict. It consists of a common biosuite that can be installed on vehicles, ships, and at fixed sites to provide biological detection and identification to all service personnel. JBPDS is portable and can support bare-base or semi-fixed sites. JBPDS will presumptively identify 10 BW agents simultaneously. It will also collect a liquid sample for confirmatory analysis and identification. Planned product improvements will focus on reducing size, weight, and power consumption while increasing system reliability.

JBPDS can operate from a local controller on the front of each system, remotely, or as a network. JBPDS meets all environmental, vibration, and shock requirements of its intended platforms, as well as requirements for reliability, availability, and maintainability. JBPDS includes both military and commercial global positioning, meteorological, and network modem capabilities. The system will interface with the Joint Warning and Reporting Network (JWARN).

Program Status

- **1QFY07:** Low rate initial production

Recent and Projected Activities

- **2QFY07:** Follow-on operational test and evaluation
- **3QFY08:** Full rate production decision

ACQUISITION PHASE

- Concept & Technology Development
- System Development & Demonstration
- **Production & Deployment**
- Operations & Support

Joint Biological Point Detection System (JBPDS)

FOREIGN MILITARY SALES
None

CONTRACTORS
General Dynamics ATP Division (Charlotte, NC)

Joint Chemical Agent Detector (JCAD)

Protects U.S. forces by detecting, identifying, quantifying, alerting, and reporting the presence of chemical warfare agents.

INVESTMENT COMPONENT

- Modernization
- Recapitalization
- Maintenance

Description & Specifications

The Joint Chemical Agent Detector (JCAD) is a detector or network of detectors capable of automatically detecting, identifying, and quantifying chemical agents inside aircraft and shipboard interiors, providing handheld monitoring capabilities, and protecting the individual Soldier, airman, and Marine through the use of pocket-sized detection and alarm.

The JCAD program will provide the services a handheld combined portable monitoring and chemical agent point detector for ship, aircraft, and individual warfighter applications. JCAD will automatically and simultaneously detect, identify, and quantify chemical agents in their vapor form. The detector will provide visual and audible indicators and will display the chemical agent class and relative hazard level dosage. The services will place the system on aircraft, vehicles, naval ships, at fixed sites, and on individuals designated to operate in a chemical threat area (CTA). The system will be capable of operating in a general chemical warfare environment, and can undergo conventional decontamination procedures by the warfighter. JCAD is designed to interface and be compatible with current and future anti-chemical, nuclear, and biological software.

The JCAD acquisition program market survey found that commercially available detectors could satisfy revised JCAD requirements. JCAD's restructured acquisition strategy will assess commercially available products to provide the most capable, mature system, at the best life-cycle cost. This strategy provides opportunities to leverage commercial developments for fielding expanded capabilities.

Program Status

- **4QFY06:** Complete operational assessments
- **1QFY07:** Complete developmental testing

Recent and Projected Activities

- **2QFY07:** Milestone C low-rate initial production decision
- **3QFY07:** Multi-service operational test and evaluation
- **4QFY07:** Full rate production decision

ACQUISITION PHASE

- Concept & Technology Development
- System Development & Demonstration
- Production & Deployment
- Operations & Support

Joint Chemical Agent Detector (JCAD)

FOREIGN MILITARY SALES
None

CONTRACTORS
Smith's Detection (Edgewood, MD)

Joint Combat Identification Marking System (JCIMS)

Improves combat effectiveness and reduces incidents of fratricide by providing a cost-effective and proven means to positively identify friendly ground forces on the battlefield.

INVESTMENT COMPONENT

- Modernization
- Recapitalization
- Maintenance

Description & Specifications

The Joint Combat Identification Marking System (JCIMS) consists of combat identification panels (CIP), thermal identification panels (TIP), and Phoenix infrared (IR) light.

CIPs are aluminum panels covered with a thermal film that produces a "cold" spot on a hot background when viewed through thermal sights. CIPs provide ground-to-ground and limited air-to-ground target identification.

TIPs are cloth panels covered with a thermal film similar to that used by CIPs. TIPs provide air-to-ground and a limited ground-to-ground target identification capability by providing a "cold" spot similar to CIPs when viewed through aircraft or vehicle thermal sights.

Phoenix Lights are IR blinking strobes visible through night vision goggles (NVG), which provide ground-to-ground and air-to-ground target identification.

Program Status

- **Current:** Production and fielding

Recent and Projected Activities

- **2QFY07–1QFY09:** Support Operation Iraqi Freedom; design and production of JCIMS for United States Marine Corps

ACQUISITION PHASE

Concept & Technology Development | System Development & Demonstration | Production & Deployment | Operations & Support

Joint Combat Identification Marking System (JCIMS)

FOREIGN MILITARY SALES
Afghanistan (recent requisitions placed)
Australia, Canada

CONTRACTORS
Crossroads Industrial Services (Indianapolis, IN)
Rauch, Inc. (New Albany, IN)
Night Vision Equipment Company (Emmaus, PA)
Eagle Industries, Inc. (St. Louis, MO)

Combat Identification Panels (CIP)

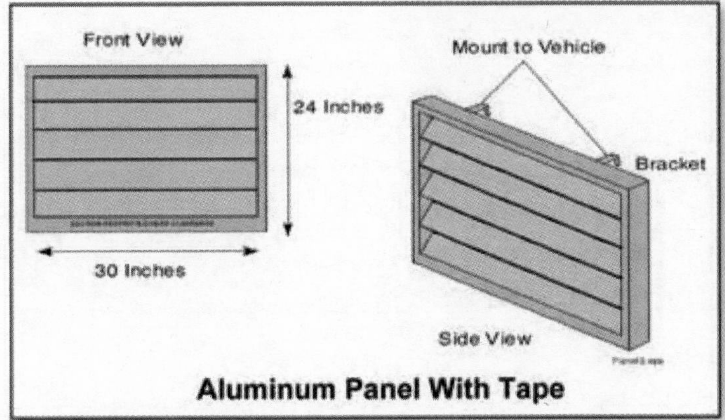

Aluminum Panel With Tape

Thermal Identification Panels (TIP)

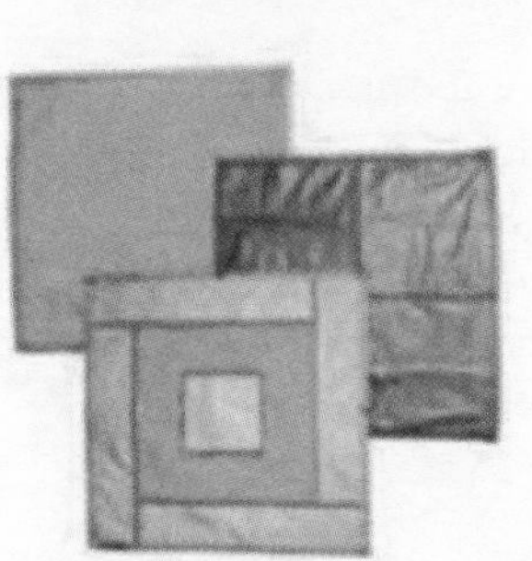

Cloth Panel With Tape

Phoenix IR Light

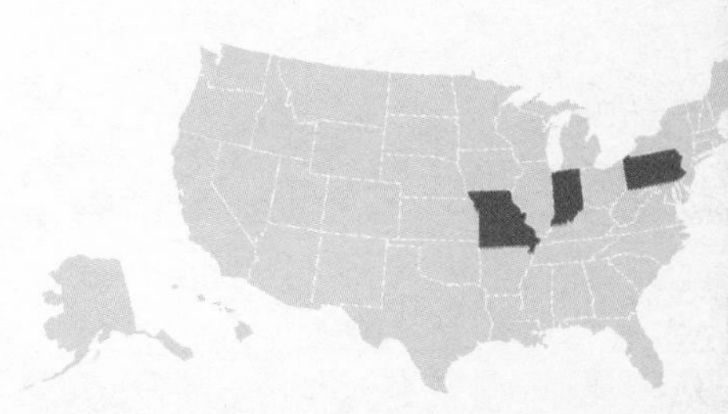

Joint Common Missile (JCM)

Improves operational effectiveness, range and lethality, and Soldier/aircraft survivability with advanced line-of-sight and beyond line-of-sight, anti-armor, and anti-materiel engagement capability and precision strike and fire-and-forget technologies.

INVESTMENT COMPONENT

Modernization

Recapitalization

Maintenance

Description & Specifications

The Joint Common Missile (JCM) will respond to expanding regional threats, joint/international operations, and missile stockpile shortages with a single missile for both air-launched (rotary and fixed-wing) and ground-launched missions, while providing flexibility during combat operation.

The JCM will be initially fielded to the U.S. Army Apache Longbow (AH-64D), the U.S. Navy Super Hornet (F/A-18E/F) and Seahawk (MH-60R), and the Marine Corps Super Cobra (AH-1Z).

JCM is designed for use on a wide variety of joint and international platforms including the United Kingdom's Harrier II Plus (AV-8B), Apache (AH-64A), Kiowa Warrior (OH-58D), Joint Strike Fighter, Multi-mission Maritime Aircraft (MMA), Armed Reconnaissance Helicopter (ARH-70A), Special Operations Forces MH-60L/M DAP and Little Bird (AH-6J/M), Unmanned Aircraft Systems (ERMP), and has potential for ground vehicles. The modular design will reduce life-cycle costs, including demilitarization, and allow for continuous technology insertion to ensure improvements against evolving threats. JCM will effectively engage and destroy a variety of targets, including stationary and moving or re-locatable, high-value threat targets, as well as bunkers and other structures on the digital battlefield, well into the future. It will be designed and tested to achieve the following:

- Fire-and-forget and precision strike
- Increased stand-off range
- Increased survivability (both missile and platform)
- Multi-purpose warhead for increased lethality (military operations in urban terrain [MOUT] structures, heavy armor, and patrol craft)
- Multi-mode seeker for increased performance given adverse weather or countermeasures
- Modularity to enable technology insertion for capability enhancement and shelf-life extension, and to facilitate demilitarization
- State-of-the-art performance in the area of Insensitive Munitions compliance

Diameter: 7 inches
Weight: 108 pounds
Length: 70 inches

Program Status

- **3QFY04:** Milestone B Defense Acquisition Board approval
- **1QFY05:** Program placed into Technology Maturation phase
- **1QFY07:** Decision expected on JCM Path Forward

Recent and Projected Activities

- Pending decision on restart of the program based on the Joint Research Oversight Council (JROC) review

ACQUISITION PHASE

Concept & Technology Development | System Development & Demonstration | Production & Deployment | Operations & Support

Joint Common Missile (JCM)

FOREIGN MILITARY SALES
None

CONTRACTORS
Lockheed Martin (Orlando, FL)
Aerojet (Sacramento, CA)
Lockheed Martin (Ocala, FL)
REMEC (San Diego, CA)
General Dynamics (Niceville, FL)

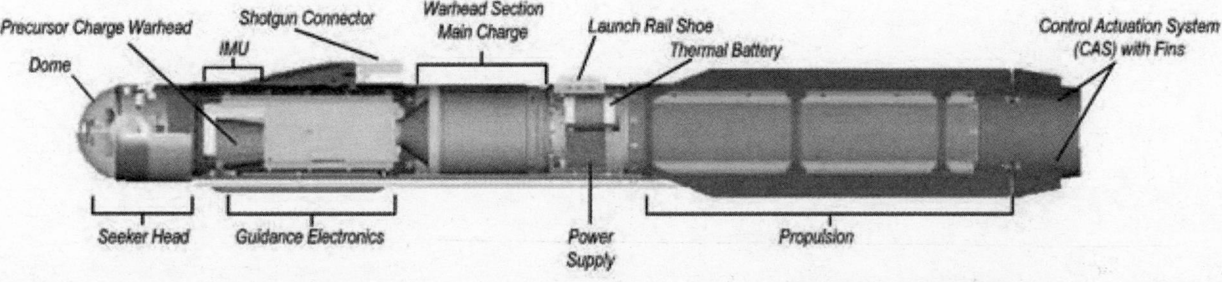

Joint High Speed Vessel (JHSV)

Provides high-speed intra-theater transport of troops and cargo.

INVESTMENT COMPONENT

Modernization

Recapitalization

Maintenance

Description & Specifications

The Joint High Speed Vessel (JHSV) is a high-speed (35 knots), shallow draft sealift platform that maximizes current commercial fast ferry technology, and represents the next-generation of Army watercraft to support the Army's doctrinal intra-theater lift mission. JHSV provides flexibility and agility within a theater, enabling the Joint Force Commander to insert combat power and sustainment into austere ports worldwide.

Supporting Army prepositioned stocks and joint logistics over-the-shore, JHSV expands the reach and possibilities of prepositioning both on land and afloat. JHSV provides the capability to conduct operational maneuver and repositioning of intact unit sets while conducting en route mission planning and rehearsal. JHSV provides the combatant commander with increased throughput, increased survivability, increased responsiveness, and improved closure rates. This transport transformation-enabler helps achieve force deployment goals and full distribution-based logistics. JHSV offers the joint force commander a multi-modal and multi-purpose platform to support joint operations that complements C-17 and C-130 airlift capabilities and minimizes the need for large-scale reception, staging, onward movement, and integration of Soldiers, vehicles, and equipment within the battlespace. The vessel will have the following additional features:

- Flight deck
- Joint interoperable, command, control, communications, computers, intelligence, surveillance, and reconnaissance (C4ISR)
- Underway refueling
- Movement tracking system
- Electronic navigation
- Integrated materiel handling

Program Status

- **3QFY06:** Army Theater Support Vessel (TSV) advanced concept technology demonstration (ACTD) completed to provide technological basis for JHSV Program
- **3QFY06:** Defense Acquisition Board (DAB) Milestone A review

Recent and Projected Activities

- **2QFY08:** DAB Milestone B review
- **2QFY08:** Detailed design and construction contract award

ACQUISITION PHASE

Concept & Technology Development | System Development & Demonstration | Production & Deployment | Operations & Support

Technical Approach

Known Commercial Designs

Austal 126
INCAT 112
TSL - 140
MDV-300
Austal 105

JOINT HIGH SPEED VESSEL
NAVY
ARMY
MARINES

Known Military Designs

LM-LCS
GD-LCS

Hybrid Designs

MONO-HULL
Semi-Swath CAT
Wave Piercing CAT
Trimaran
Surface Effect Ship

Advanced Strategic/Operational Lift – Joint High Speed Vessel (JHSV)

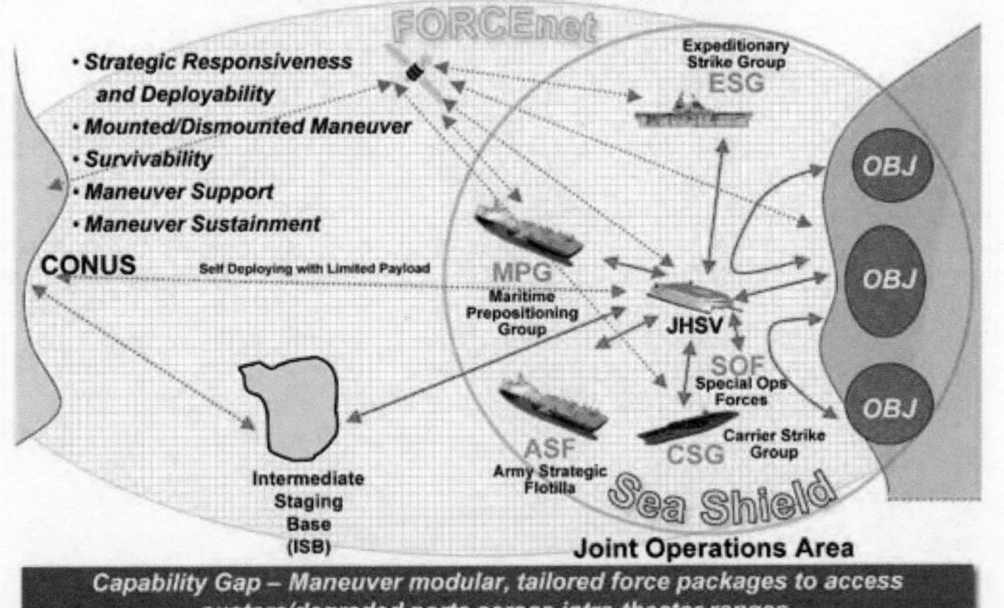

Joint High Speed Vessel (JHSV)

FOREIGN MILITARY SALES
None

CONTRACTORS
To be selected

Joint Land Attack Cruise Missile Defense (LACMD) Elevated Netted Sensor System (JLENS)

Provides over-the horizon detection, tracking, and classification of cruise missiles and other air targets so they can be engaged by air-directed, surface-to-air missiles or air-directed, air-to-air missile defense systems.

INVESTMENT COMPONENT

- Modernization
- Recapitalization
- Maintenance

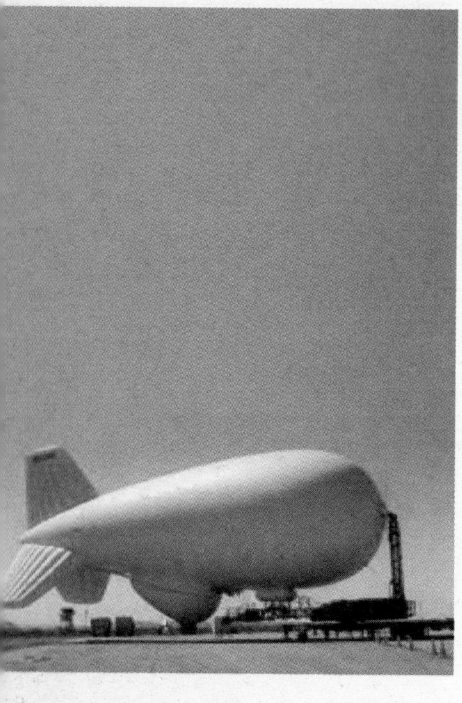

Description & Specifications

The Joint Land Attack Cruise Missile Defense (LACMD) Elevated Netted Sensor System (JLENS) is a joint interest program that protects U.S., allied, and coalition forces, civilian population centers, and critical military and geopolitical assets from air and missile attacks. JLENS is a crucial part of the Integrated Air and Missile Defense (IAMD) architecture that will counter Land Attack Cruise Missiles (LACMs) and low flying aerial threats. JLENS targets not only LACMs but also unmanned aerial vehicles, unmanned combat aerial vehicles, rotary-wing and fixed-wing aircraft, theater ballistic missiles, large caliber rockets, and surface moving targets.

A JLENS Orbit consists of two systems: a fire control radar system and a wide-area surveillance radar system. Each radar system employs a separate 74-meter tethered aerostat, mobile mooring station, radar and communications payload, mobile processing station, and associated ground support equipment. A JLENS system consists of either the fire control radar or the wide-area surveillance radar with associated platform and ground equipment.

The fire control and wide-area surveillance radars have the capability of operating autonomously and/or operating jointly as a JLENS orbit. JLENS uses advanced sensor and networking technologies to provide 360-degree, wide-area surveillance and precision tracking of land-attack cruise missiles. JLENS surveillance provides a long-range air picture enhanced by identification of friend or foe. This information contributes to the single integrated air picture.

JLENS can detecting and track surface moving targets, detect tactical ballistic missiles at boost phase, and large caliber rockets during ascent phase. As a multi-role platform, JLENS enables extended range command and control linkages, communications relay, and battlefield situational awareness. JLENS is the only elevated persistent long-range surveillance and integrated fire control sensor in the IADM architecture.

Program Status

- **4QFY05:** Milestone B approval to enter system development and demonstration (SDD) phase
- **1QFY06:** Change order modification to JLENS contract to begin SDD
- **2QFY06:** JLENS weapon system review
- **4QFY06:** SDD definitization proposal received
- **4QFY06:** JLENS Orbit system requirements review
- **4QFY06–1QFY07:** SDD proposal evaluation
- **1QFY07:** Complete SDD contract definitization
- **1QFY07:** Conduct JLENS Orbit system functional review

Recent and Projected Activities

- **2QFY07:** Conduct JLENS Spiral 2 SDD Integrated Baseline Review
- **4QFY07:** Conduct JLENS Orbit preliminary design review
- **4QFY08:** Conduct JLENS Orbit critical design review

ACQUISITION PHASE

Concept & Technology Development | System Development & Demonstration | Production & Deployment | Operations & Support

Joint Land Attack Cruise Missile Defense (LACMD) Elevated Netted Sensor System (JLENS)

FOREIGN MILITARY SALES
None

CONTRACTORS
System development and demonstration:
Raytheon (Andover, MA; El Segundo, CA)
TCOM (Columbia, MD)
System Engineering Technical Analysis (SETA) support:
CAS Inc. (Huntsville, AL)

Joint Land Component Constructive Training Capability (JLCCTC)

Provides tools to train unit commanders and their staffs from battalion through theater levels.

INVESTMENT COMPONENT
- Modernization
- Recapitalization
- Maintenance

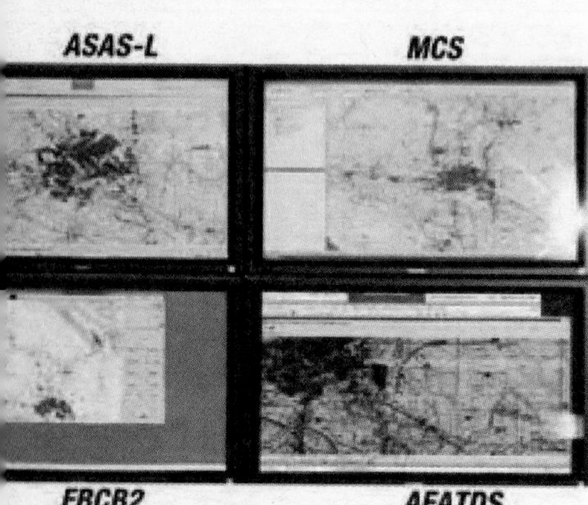

Description & Specifications

The Joint Land Component Constructive Training Capability (JLCCTC) is a collection of integrated simulation models that help Army Battle Command Systems (ABCS) facilitate command and staff training. These federated systems include Warfighters Simulation (WARSIM), One Semi-Automated Forces (OneSAF), Digital Battle Staff Trainer ((DBST), Corps Battle Simulation, and Tactical Simulation (TACSIM).

JLCCTC-Multi-Resolution Federation (MRF) is a command post exercise driver that supports training of commanders and their staffs in maneuver, logistics, intelligence, air defense, and artillery. It is a set of constructive simulation software, supporting software, and commerical off-the-shelf hardware. The components are connected by a combination of the standard High Level Architecture Run Time Infrastructure, Distributed Interactive Simulation (DIS), custom interfaces to the Corps Battle Staff Master Interface and Point-to-Point. The JLCCTC provides a simulated operational environment, in which computer generated forces stimulate and respond to the command and control processes of the commanders and staffs. The JLCCTC models will provide full training functionality for leader and battle staff for the Army and the joint, interagency, intergovernmental, and multinational spectrum. The JLCCTC provides an interface to Operational Battle Command equipment allowing commanders and their staffs to train with their "go to war" systems.

Joint Land Component Constructive Training Capability-Entity Resolution Federation (JLCCTC-ERF) is a federation of simulations, simulation C4I interfaces, data collection, and after action review tools. It stimulates the Army Battle Command System (ABCS) to facilitate battle staff collective training by requiring staff reaction to incoming digital information while executing the commander's tactical plan. The targeted training audience is brigade and battalion battle staffs, including functional command post training and full command post training. Battle staffs above brigade echelons may also employ JLCCTC-ERF to achieve specific training objectives.

Program Status

- **FY06:** Continued to improve system performance and increase scalability
- **FY06:** Integrated training system components
- **4QFY06:** JLCCTC Version 3 fielding
- **1QFY07:** Milestone C production and deployment

Recent and Projected Activities

- **1QFY07–3QFY07:** Design, testing, integration, and validation (V. 4)
- **4QFY07:** JLCCTC Version 4 release
- **1QFY08–3QFY08:** Design, testing, integration and validation (V. 5)
- **4QFY08:** JLCCTC V. 5 release

ACQUISITION PHASE

Concept & Technology Development | System Development & Demonstration | Production & Deployment | Operations & Support

Joint Land Component Constructive Training Capability (JLCCTC)

FOREIGN MILITARY SALES
None

CONTRACTORS
Lockheed Martin Information Systems (Orlando, FL)
Tapestry Solutions (San Diego, CA)
The Aegis Technology Group, inc. (Orlando, FL)

Commanders Battlefield Visualization Display Projected From BattleSight Projectors

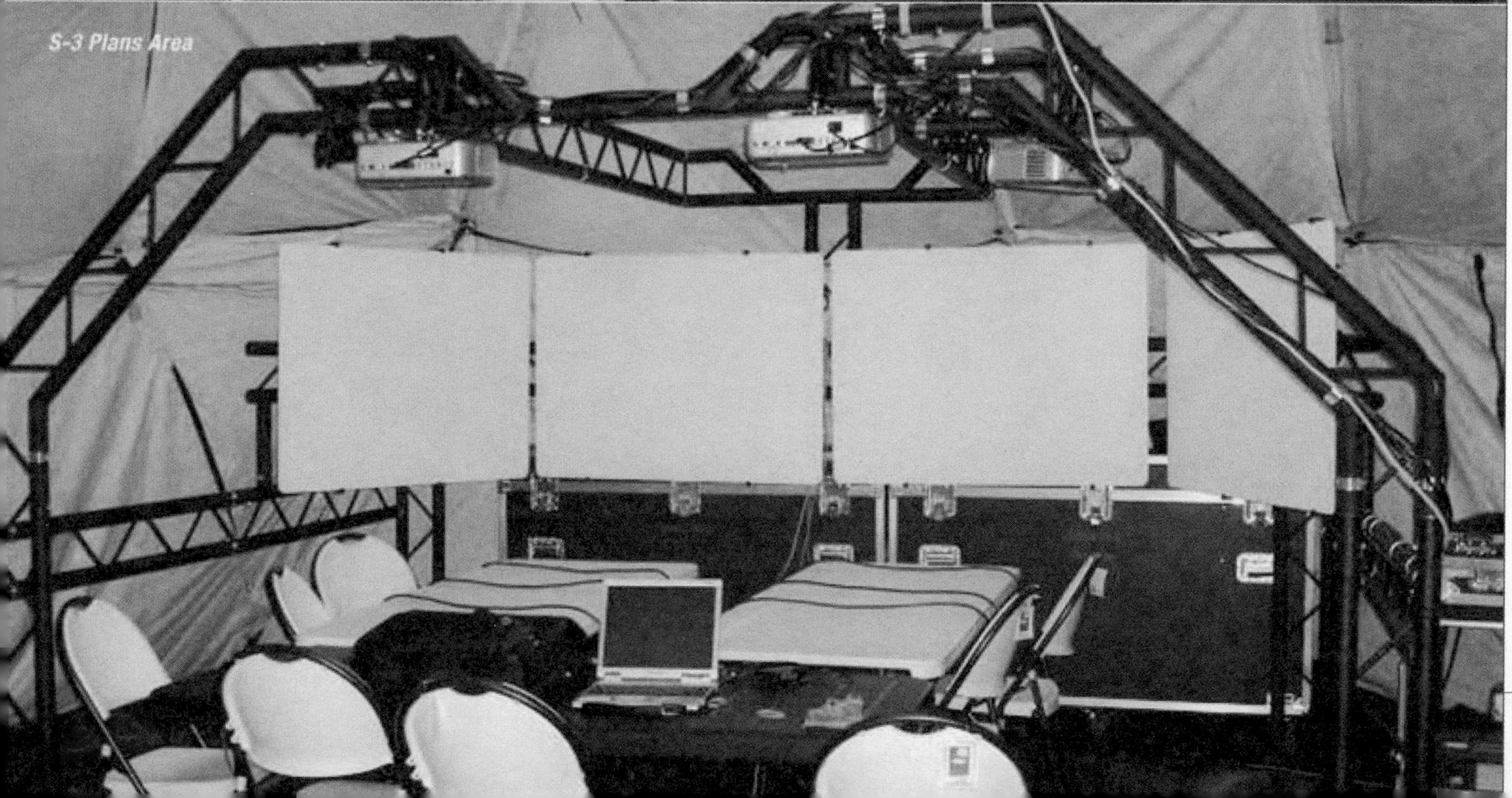

S-3 Plans Area

Joint Network Management Systems (JNMS)

Provides a common, automated planning and management tool that will plan, monitor, and control the joint communications and data backbone associated with a joint task force or joint special operations task force.

INVESTMENT COMPONENT

Modernization

Recapitalization

Maintenance

Description & Specifications

The Joint Network Management Systems (JNMS) is an automated software system that will provide communications planners with a common set of tools to conduct high-level planning, detailed planning and engineering, monitoring, control and reconfiguration, spectrum planning and management, and security of systems. JNMS consists primarily of commercial off-the-shelf (COTS) software modules that provide capabilities to accomplish its mission.

JNMS will be developed and implemented in increments based on incorporating key performance parameter (KPP) threshold requirements, non-KPP threshold requirements, and objective requirements.

The JNMS program has been modified to provide more modularized network planning, network management, and trouble ticketing capabilities that can be tailored to each fielding site's unique requirements. This is done by specifically splitting these functionalities into separate laptop-based sub-systems, thus significantly reducing the original system footprint.

Program Status

- **1QFY06:** Full rate production approved
- **3QFY06:** "New Way Ahead" re-defined JNMS system composition
- **4QFY06:** Materiel release

Recent and Projected Activities

- **2QFY07:** Fieldings re-commence
- **2QFY07:** S/W Release 1.4
- **4QFY07:** S/W Release 1.5.1
- **4QFY08:** S/W Release 1.5.2

ACQUISITION PHASE

Concept & Technology Development | System Development & Demonstration | Production & Deployment | Operations & Support

Joint Network Management Systems (JNMS)

FOREIGN MILITARY SALES
None

CONTRACTORS
SAIC (San Diego, CA; Piscataway, NJ)

Joint Precision Airdrop System (JPADS)

Provides the warfighter with precision airdrop ensuring accurate delivery of supplies to forward operating forces, reducing vehicular convoys, and allowing aircraft to drop cargo at safer altitudes and off-set distances.

INVESTMENT COMPONENT

Modernization

Recapitalization

Maintenance

Description & Specifications

The Joint Precision Airdrop System (JPADS) integrates a cargo pallet, cargo net, tactical parachute, and precision guidance interface to create a system which can accurately drop air delivered supplies. The system is being developed in four weight classes; 2,000 pounds, 10,000 pounds, 30,000 pounds, and an objective system of 60,000 pounds. The guidance system uses global positioning and interfaces with a Mission Planning Module on board the aircraft to receive real-time weather data and compute aerial release points.

JPADS is being designed for aircraft to drop cargo from altitudes of up to 24,500 feet mean sea level. It will release cargo from a minimum off-set of 5 KM from the intended point of impact, with an objective capability of 25 KM off-set. This off-set allows aircraft to stay out of range of many anti-aircraft systems. It also allows for aircraft to drop systems from a single aerial release point and deliver them to multiple locations. Once on the ground, the precise placement of the loads greatly reduces the time needed to recover the load. In turn, exposure to ground forces is minimized as well.

Program Status:

- **1QFY07:** Approval of the capabilities development document for the 2,000- and 10,000-pound variants
- **1QFY07:** Milestone B (permission to enter system development and demonstration) for the 2,000-pound variant
- **1QFY07:** Request for proposal for the 2,000-pound variant development phase

Recent and Projected Activities:

- **2QFY07:** Award contract for 2,000-pound variant development
- **2QFY07:** 10,000 pounds variant transfers from Advanced Concept Technology Demonstration to product manager's management
- **4QFY07:** Developmental testing of the 2,000-pound variant
- **2QFY07–3QFY08:** Developmental testing of the 10,000-pound variant
- **3QFY08:** Milestone C (entrance into Production Phase) for the 2,000-pound variant

ACQUISITION PHASE

Concept & Technology Development | System Development & Demonstration | Production & Deployment | Operations & Support

Joint Precision Airdrop System (JPADS)

FOREIGN MILITARY SALES
None

CONTRACTORS
MMIST (Ontario, Canada)
Strong Enterprises (Orlando, FL)
Capewell (South Windsor, CT)

Joint Service General Purpose Mask (JSGPM)

Provides face, eye, and respiratory protection from battlefield concentrations of chemical and biological (CB) agents, toxins, toxic industrial materials, and radioactive particulate matter.

INVESTMENT COMPONENT

Modernization

Recapitalization

Maintenance

Description & Specifications

The Joint Service General Purpose Mask (JSGPM) is an ACAT III program. The JSGPM will fill a need for a lightweight, CBRN protective mask system incorporating state of the art technology to protect U.S. forces from actual or anticipated threats. The JSGPM will provide above-the-neck, head and eye/respiratory protection against chemical and biological (CB) agents, radioactive particles, and toxic industrial materials (TIMs). The mask component designs will be optimized to minimize their impact on the wearer's performance and to maximize its ability to interface with current and future service equipment and protective clothing. The JSGPM mask system is being developed to replace the M40/M42 series of masks for U.S. Army and U.S. Marine ground and combat vehicle operations and the MCU-2/P series for U.S. Air Force and U.S. Navy ground and shipboard applications.

Program Status

- **1Q–4QFY06:** Continued production of low rate initial production systems
- **1Q–4QFY06:** Continued developmental testing
- **3Q–4QFY06:** Complete multi-service operational test and evaluation

Recent and Projected Activities

- **FY07:** Full rate production
- **FY07–09:** Procure and field systems

ACQUISITION PHASE

Concept & Technology Development | System Development & Demonstration | Production & Deployment | Operations & Support

Joint Service General Purpose Mask (JSGPM)

FOREIGN MILITARY SALES
None

CONTRACTORS
Avon Protection Systems (Cadillac, MI)

Joint Tactical Ground Station (JTAGS)

Disseminates early warning, alerting, and cueing information on theater ballistic missiles and other tactical events by receiving and processing in-theater, direct, downlinked data from Defense Support Program (DSP) satellites.

INVESTMENT COMPONENT

Modernization

Recapitalization

Maintenance

Description & Specifications

The Joint Tactical Ground Station (JTAGS) supports simultaneous operations in multiple theaters and provides real-time, space-based information to theater combatant commanders.

The JTAGS processor and communications equipment is housed in an 8 x 8 x 20-foot shelter with three external high-gain antennas that receive infrared (IR) data from three satellites. The system is transportable by C-141 and larger aircraft and can operational within hours. During crisis situations, the system deploys in pairs for enhanced reliability.

The current JTAGS system has been approved for a Preplanned Product Improvement (P3I) program that envisions an enhanced mobile missile warning and communications system that will receive and process both residual direct downlinked data from DSP satellites and the new Space-Based Infrared System (SBIRS) sensors. The JTAGS successor system will also provide battlespace characterization data for enhanced situational awareness.

The Army plans to replace the five fielded JTAGS with the P3I systems beginning in FY12-13. The transition to the follow-on system is expected to occur after the SBIRS geosynchronous satellites are launched and operational. A second block improvement expected to begin in FY13 incorporates data from the technologies developed by the Missile Defense Agency (MDA) and the Space Tracking and Surveillance System (STSS) (formerly SBIRS Low).

Program Status

- **3QFY06:** Completed test and certification of LINK 16
- **4QFY06:** Begin development of JTAGS Software Test Environment (JSTE) and Secure Telephone Unit / Equipment (STU/STE) into JTAGS
- **1QFY07:** Complete integration of Joint Tactical Terminal (JTT) into JTAGS units
- **1QFY07:** Complete RT Logic Versa Module Euro (VME) integration into JTAGS units
- **1QFY07:** Begin Interim Highly Elliptical Orbit (IHEO) and Common Data Link Interface (CDLI/SIPRNET) development for JTAGS
- **1QFY07:** Integration of MIDS into JTAGS
- **1QFY07:** Initiate JTAGS P3I program

Recent and Projected Activities

- **2QFY07:** Decomposition of requirements to contract specification complete
- **2QFY07:** Continue development of IHEO, CDLI/SIPRNET, and MIDS integration into JTAGS
- **4QFY07:** Complete integration of MIDS, JSTE, and STU/STE equipment into JTAGS
- **1QFY08:** Initiate TMP system development and demonstration
- **1QFY08:** Complete integration of IHEO and CDLI/SIPRNET
- **1QFY09:** Continue development of JTAGS follow-on capability

ACQUISITION PHASE

Concept & Technology Development | System Development & Demonstration | Production & Deployment | Operations & Support

Joint Tactical Ground Stations (JTAGS)

FOREIGN MILITARY SALES
None

CONTRACTORS

Deployment, Production, and P3I Phase I:
GenCorp (Azusa, CA; Colorado Springs, CO)
System Engineering Technical Analysis (SETA) support:
Mevatec (Huntsville, AL)
Operations and support:
Northrop Grumman (Azusa, CA)
JTAGS P3I:
Lockheed Martin (Sunnyvale, CA; Boulder, CO)
Northrop Grumman (Azusa, CA)
SETA support:
BAE Systems (Huntsville, AL)

Joint Warning and Reporting Network (JWARN)

Accelerates the warfighter's response to a nuclear, biological, or chemical attack by providing joint forces the capability to report, analyze, and disseminate detection, identification, location, and warning information.

INVESTMENT COMPONENT

- Modernization
- Recapitalization
- Maintenance

Description & Specifications

The Joint Warning and Reporting Network (JWARN) is a computer-based system designed to collect analyze, identify, locate, and report information on nuclear, biological, or chemical (NBC) activity and threats from sensors in the field and to disseminate that information to decision-makers throughout the command. Located in command and control centers, JWARN will be compatible and integrated with joint service command, control, communications, computers, intelligence, and surveillance reconnaissance (C4ISR) systems. JWARN's component interface device connects to the sensors, which can detect various types of attack.

JWARN is being developed for deployment with NBC sensors in the following battlefield applications: combat and armored vehicles, tactical vehicles, vans, shelters, shipboard application, area warning, semi-fixed sites, and fixed sites. The component device relays warnings to C4ISR systems via advanced wired or wireless networks. JWARN reduces the time from incident observation to warning to less than two minutes, enhances warfighters' situational awareness throughout the area of operations, and supports battle management tasks.

The JWARN-Full Capability System will be developed as a single increment. The development phase will be followed by a preplanned product improvement effort, which will include artificial intelligence modules for NBC operations, an upgrade to match future C4ISR systems, and standard interfaces for use with future detectors.

Program Status

- **Current–1QFY07:** JCID wireless development and GCCS development and testing

Recent and Projected Activities

- **2QFY07–4QFY07:** JWARN developmental testing and operational assessment
- **2QFY08:** JWARN full rate production

ACQUISITION PHASE

- Concept & Technology Development
- System Development & Demonstration
- Production & Deployment
- Operations & Support

Joint Warning and Reporting Network (JWARN)

FOREIGN MILITARY SALES
None

CONTRACTORS
Bruhn NewTech (Ellicott City, MD)
Northrop Grumman Information Technology (Winter Park, FL)

Kiowa Warrior

Supports combat and contingency operations with a light, rapidly deployable helicopter capable of armed reconnaissance, security, target acquisition and designation, command and control, light attack, and defensive air combat missions.

INVESTMENT COMPONENT

Modernization

Recapitalization

Maintenance

Description & Specifications

The Kiowa Warrior is a rapidly deployable single-engine, two-man, lightly armed reconnaissance helicopter that features advanced visionics, navigation, communication, weapons, and cockpit integration systems. Its mast-mounted sight houses a thermal imaging system, low-light television, and a laser rangefinder/designator permitting target acquisition and engagement at stand-off ranges and in adverse weather. The navigation system can convey precise target locations to other aircraft or artillery via its advanced digital communications system. It can also transmit battlefield imagery to deliver near real-time situational awareness to command and control elements.

The Kiowa Warrior provides anti-armor, anti-personnel, and anti-aircraft capabilities at stand-off ranges. Kiowa Warrior fielding is complete, but the Army is installing safety and performance modifications to keep the aircraft safe and mission-effective until it is retired.

Program Status

- **2QFY06:** Contract for SEP Lot IX (18 aircraft) awarded
- **2QFY06:** Completed transformation of 2/6 Cavalry Squadron, Hawaii
- **3QFY06:** Completed transformation of 4/6 Armored Cavalry Squadron, Ft. Lewis
- **3QFY06:** Completed installation of engine barrier filters on entire fleet
- **4QFY06:** Completed transformation of 7/17 Cavalry Squadron, Ft. Campbell
- **4QFY06:** Completed conversion of Category B Trainers to Safety Enhancement Configuration
- **FY05–FY06:** Reset of 131 aircraft completed
- **FY05–FY06:** Supported OIF, six units involved; highest level was 107 aircraft, 60 percent of operational fleet

Recent and Projected Activities

- **2QFY07:** Complete SEP Lot VIII deliveries
- **FY07–FY09:** Reset/Operation Iraqi Freedom support
- **FY07:** Begin weight reduction program modification
- **1QFY08:** Complete transformation of 6/17 Cavalry Squadron, AK
- **2QFY08:** Complete transformation of 1/6 Cavalry Squadron, Ft. Carson
- **2QFY08:** Complete SEP Lot IX deliveries
- **3QFY08:** Complete SEP fielding to 1/230th, ARNG, Knoxville, TN
- **FY09:** Begin retirement of Kiowa Warrior as the Army fields the Armed Reconnaissance Helicopter

ACQUISITION PHASE

Concept & Technology Development | System Development & Demonstration | Production & Deployment | Operations & Support

Kiowa Warrior

FOREIGN MILITARY SALES
Delivery of 39 Kiowa Warriors to Taiwan completed

CONTRACTORS
Bell Helicopter, Textron (Fort Worth, TX)
DRS Optronics Inc. (Palm Bay, FL)
Rolls Royce Corp. (Indianapolis, IN)
Honeywell, Inc. (Albuquerque, NM)
Smiths (Grand Rapids, MI)

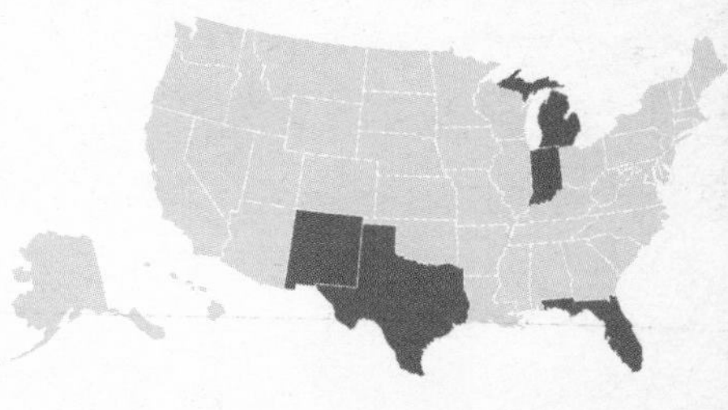

Land Warrior (LW)

Provides unprecedented tactical awareness and significant improvements in lethality, survivability, mobility, and sustainment to dismounted Soldiers and units engaged in the close fight.

INVESTMENT COMPONENT

Modernization

Recapitalization

Maintenance

Description & Specifications

Land Warrior (LW) is a first-generation, integrated, modular fighting system that uses state-of-the-art computer, communications, and geolocation technologies to link dismounted Soldiers to the digital battlefield. LW system components include a computer, helmet-mounted display (HMD), navigation module, voice/data radio, and a multi-functional laser that are integrated with the Soldier's mission equipment. The integrated, modular system approach optimizes fightability with minimal impact on the Soldier's combat load and logistical footprint. LW is interoperable with the Army Battle Command System.

LW-equipped Soldiers are capable of instant voice and data communications with other Soldiers, command posts, and supporting vehicles and aircraft. With his HMD, a LW-equipped Soldier can see his location, the location of other LW-equipped Soldiers, known enemy positions, and operational graphics on a large-scale map display. He can also engage the enemy, exposing only his hands, by using HMD target images from the weapon-mounted video gunsight or thermal sight. With the multi-functional laser, a Soldier can designate an enemy position to appear on the common map situational awareness display. He can also transmit the enemy position instantly to fire support elements in a digital call-for-fire.

Team Soldier equipped the 4th Battalion, 9th Infantry, 4th Stryker Brigade Combat Team, 2nd Infantry Division at Ft. Lewis, WA, with LW and Mounted Warrior (MW) to conduct a comprehensive assessment across the areas of Doctrine, Organizations, Training, Materiel, Leadership and Education, Personnel, and Facilities (DOTMLPF) and Tactics, Techniques, and Procedures (TTP). Results of the assessment will be used to determine Basis of Issue (BOI) and inform future Soldier system development decisions. As a result of positive preliminary results, the assessment unit leadership decided to keep the LW and MW systems for training and eventual deployment to combat.

Program Status

- **2QFY05:** Vice Chief of Staff of the Army approved equipping one Stryker Battalion with LW and MW for assessment
- **3QFY06:** LW/MW NET training at Ft. Lewis
- **3QFY06–1QFY07:** LW/MW DOTMLPF assessment and limited user test

Recent and Projected Activities

- **2QFY07:** Milestone C Decision
- **2QFY08:** Initial operational test and evaluation
- **3QFY08:** Full rate production

ACQUISITION PHASE

Concept & Technology Development | System Development & Demonstration | Production & Deployment | Operations & Support

Land Warrior (LW)

FOREIGN MILITARY SALES

LW has been demonstrated to more than 40 foreign countries, many of which (including the United Kingdom, Australia, Canada, and the Netherlands) have expressed continued interest in LW technology. Panel III NATO has approved the NATO Soldier Modernization Plan, which includes a requirement for the LW.

CONTRACTORS

Prime:

General Dynamics C4 Systems (Scottsdale, AZ)
Computer Sciences Corp. (Eatontown, NJ)
Omega Training Group (Columbus, GA)
Raytheon (El Segundo, CA)
General Dynamics C4 Systems (Taunton, MA)
General Dynamics Land Systems (Sterling Hts, MI)

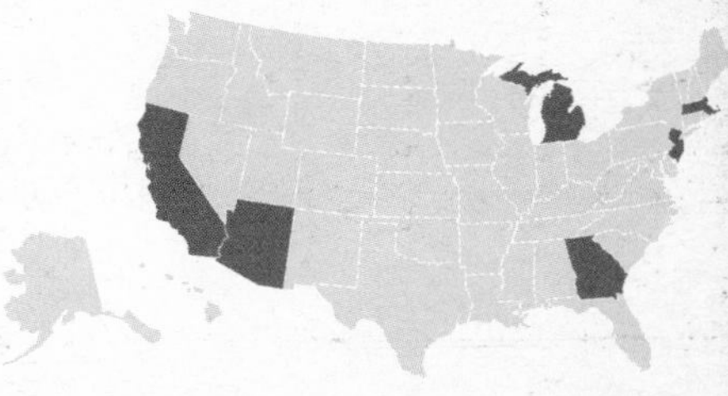

Light Utility Helicopter (LUH)

Conducts a variety of missions encompassing homeland security, civil search and rescue operations, damage assessment, test and training center support, MEDEVAC missions, and Continental United States (CONUS) counterdrug operations.

INVESTMENT COMPONENT

Modernization

Recapitalization

Maintenance

Description & Specifications

The UH-72A Light Utility Helicopter (LUH) was procured as an FAA type certified Commercial/Non-Developmental Item (NDI) single pilot, IRF certified aircraft and entered the Army Acquisition timeline at Milestone C. The UH-72A aircraft will replace the aging OH-58 and UH-1 aircraft fleets and will allow tactical Blackhawk (UH-60) aircraft to be cascaded to Army National Guard (ARNG) units in support of critical wartime requirements. The aircraft will have NVG, day, night, IFR, VFR and external hoist capabilities. Each aircraft will be equipped with wire strike protection, crashworthy fuel systems and crashworthy seats in accordance with Federal Aviation regulations. The aircraft will be fully interoperable with civil, military and government agencies. The MEDEVAC version will be able to carry two NATO standard litters and one attendant. The primary users of the LUH are Active Army Table of Distribution and Allowances (TDA) units and ARNG units. Three hundred twenty-two UH-72A aircraft will be procured and will be deployable only to permissive environments.

Max gross weight: 7,903 (pounds)
Cruise speed: 142 (knots)
Max range: 303 (nautical miles)
Endurance: 3.3 (hours)
Engines (2 each): Turbomeca Arriel 1E2
External load: 1,107 @ 4K ft/90 deg F (pounds)
Internal load: 6/1,107 @ 4K ft/90 deg F (passengers/pounds)
Crew: 11

Program Status

- **3QFY05:** Concept development document approved
- **1QFY06:** Source selection activities began
- **2QFY06** LUH Program delegated to the Army by OSD
- **3QFY06** Milestone C/low rate initial production (LRIP) decision made; program designated ACAT 1C
- **3QFY06:** Contract awarded to EADS North America for an initial 8 aircraft
- **1QFY07:** Contract modified for an additional 34 aircraft to complete LRIP
- **1QFY07:** First aircraft delivered

Recent and Projected Activities

- **2QFY07:** Conduct initial operational test
- **3QFY07:** First unit equipped; award full rate production decision

ACQUISITION PHASE

Concept & Technology Development | System Development & Demonstration | Production & Deployment | Operations & Support

Light Utility Helicopter (LUH)

FOREIGN MILITARY SALES
None

CONTRACTORS
EADS North America (Arlington, VA)
American Eurocopter (Columbus, MS; Grand Prairie, TX)
CAE USA (Tampa, FL)
Sikorsky (Stratford, CT)
Westwind Technologies (Huntsville, AL)

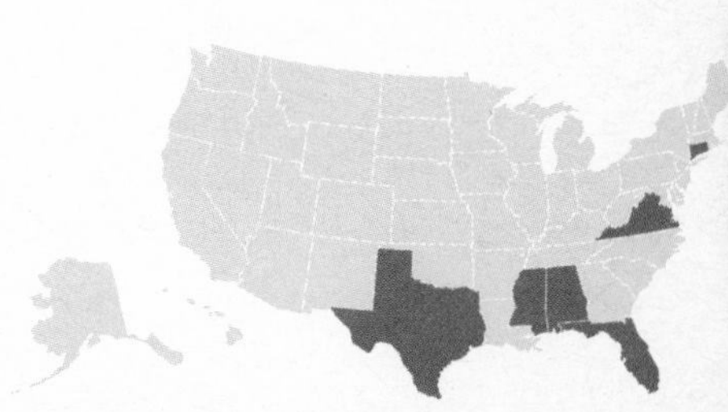

Lightweight 155mm Howitzer (LW 155)

Provides direct, reinforcing, and general support fires to maneuver forces as well as direct support artillery for Stryker Brigade Combat Teams.

INVESTMENT COMPONENT

Modernization

Recapitalization

Maintenance

Description & Specifications

The Lightweight 155mm Howitzer (M777) is the general support artillery for the Army's light forces. The use of titanium in its major structures makes it 7,000 pounds lighter than its predecessor (the M198) with no sacrifice in range, stability, accuracy, or durability. Two M777s can be transported by a C-130, and it can be dropped by parachute. The M777's lighter weight, smaller footprint, and lower profile increase strategic deployability, tactical mobility, and survivability. The automatic primer feeding mechanism, loader-assist, digital fire control, and other improvements enhance reliability and give light artillery a semi-autonomous capability found only in self-propelled howitzers.

The M777 is jointly managed, with the Marine Corps having led the development of the howitzer and the Army having led the development of Towed Artillery Digitization (TAD), the digital fire control system for the M777. Once type classified, the digital fire control-equipped howitzer will be designated the M777A1.

Software updates and the Platform Integration Kit (PIK) hardware will give the M777A2 the capability to fire the Excalibur precision guided munition. The specifications of the Excalibur-compatible howitzer are:

- Weight: 10,000 pounds or lighter with TAD
- Emplace: Less than three minutes
- Displace: Two to three minutes
- Maximum range: 30 km (assisted)
- Rate-of-fire: Four to eight rounds per minute maximum; two rounds per minute sustained
- Ground mobility: Family of Medium Tactical Vehicles, Medium Tactical Vehicle Replacement, five-ton trucks
- Air mobility: Two per C-130; six per C-17; 12 per C-5; CH-53D/E; CH-47D; MV-22
- 155mm compatibility: all fielded and developing NATO munitions
- Digital fire control: self-locating and pointing; digital and voice communications; self-contained power supply.

Program Status

- **Current:** Completed low rate initial production (LRIP) for 94 Marine Corps howitzers with conventional fire control; joint full rate production of 495 M777A1 systems ongoing; first units fielded; digital fire control program synchronized with basic howitzer program; future weapons will use digital fire control, with optical fire control as backup; previous built LRIP howitzers retrofitted with digital fire control

Recent and Projected Activities

- **2QFY07:** Army initial operational capability of M777A1 (howitzer with digital fire control)
- **1QFY10:** Full operational capability

ACQUISITION PHASE

Concept & Technology Development | System Development & Demonstration | Production & Deployment | Operations & Support

Lightweight 155mm Howitzer (LW 155)

FOREIGN MILITARY SALES

The LW 155 development was a cooperative effort with both the United Kingdom (UK) and Italy, and the UK remains a partner in production. To date, Canada has purchased 12 howitzers under an FMS case with six delivered and four currently deployed to Afghanistan. The UK and Australia are likely to purchase the system in the near future.

CONTRACTORS

Prime:
BAE Systems (United Kingdom and Hattiesburg, MS)
Castings:
Precision Castparts Corp. (Portland, OR)
Howmet Castings (Whitehall, MI)
Digital Fire Control:
General Dynamics (Burlington, VT)
Howitzer body:
Triumph Systems Los Angeles (Chatsworth, CA)

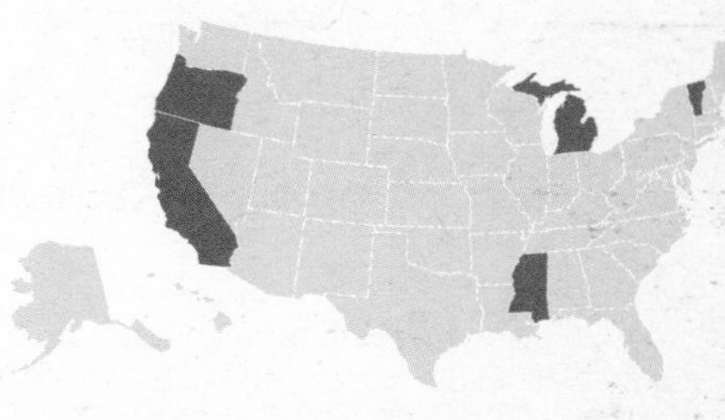

Lightweight Laser Designator Rangefinder (LLDR)

Provides fire support teams and forward observers with a man-portable capability to observe and accurately locate targets, digitally transmit target location data to the tactical network, and laser-designate high-priority targets for destruction.

INVESTMENT COMPONENT

Modernization

Recapitalization

Maintenance

Description & Specifications

The Lightweight Laser Designator Rangefinder (LLDR) is a man-portable, modular, target location and laser designation system. The two primary components are the target locator module (TLM) and the laser designator module (LDM). The TLM can be used as a stand-alone device or in conjunction with the LDM. Total system weight to conduct a 24-hour mission is 35 pounds.

The TLM incorporates a thermal imager, day camera, electronic display, eye-safe laser rangefinder, digital magnetic compass, global positioning system (GPS) electronics, and digital export capability. The TLM has an integral capability for boresighting with the LDM, allowing the operator to see the laser spot and align the system. At night and in obscured battlefield conditions, the operator can recognize vehicle-sized targets at greater than 2.5 kilometers. During day operations, the operator can recognize targets at a distance of greater than 7 kilometers. At a range of 10 kilometers, the operarator can locate targets to less than 40 meters. The LDM emits coded laser pulses compatible with Department of Defense and NATO laser-guided munitions. Users can designate targets at ranges greater than 5 kilometers.

Program Status

- **Current:** In full-rate production
- **1QFY07:** Full materiel release
- **1QFY07:** Award of Preplanned Product Improvement contract for improved day-TV camera, obsolescence reset, and weight reduction

Recent and Projected Activities

- Continue fielding

ACQUISITION PHASE

Concept & Technology Development | System Development & Demonstration | **Production & Deployment** | Operations & Support

Lightweight Laser Designator Rangefinder (LLDR)

FOREIGN MILITARY SALES
None

CONTRACTORS
Prime:
Northrop Grumman Electronic Systems Laser Systems Division (Apopka, FL)
Thermal Imager:
CMC Electronics, Cincinnati (Mason, OH)
Thermal Imager:
Indigo Systems (Santa Barbara, CA)

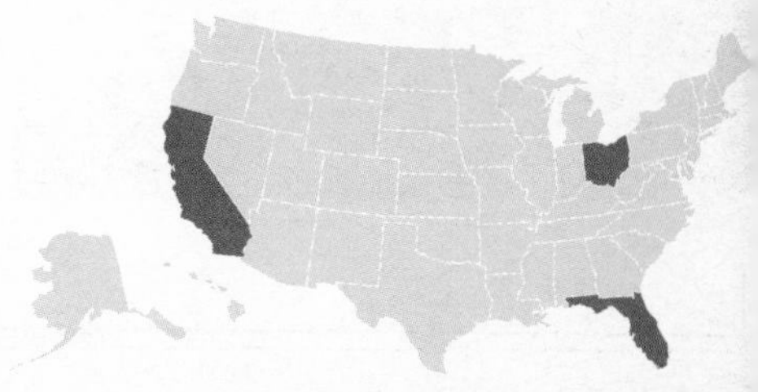

Lightweight Water Purification (LWP) System

Provides a safe, reliable, supply of potable water to support ground and amphibious troops, Special Operation Forces, and air mobile/airborne units.

INVESTMENT COMPONENT

- **Modernization**
- Recapitalization
- Maintenance

Description & Specifications

The Lightweight Water Purification (LWP) system consists of a feed water pump, hoses, Reverse Osmosis Water Purification Unit (ROWPU) elements, ultra filtration, high-pressure pump, control panel, 3-kilowatt tactical quiet generator, and a 1,000-gallon water storage and distribution system. The LWP can purify chemical, biological, and nuclear contaminated water. It is mounted on modular skids that can be lifted by four people and will normally be transported in a two-soldier cargo High Mobility Multi-Purpose Wheeled Vehicle (HMMWV). The LWP can be sling transported by utility helicopter (UH-60) and is air droppable from C-17/C-130 fixed-wing aircraft. It is capable of supplying 125 gallons per hour (GPH) of potable water from a fresh water source and 75 GPH from a salt water source to support a wide range of tactical and contingency operations.

Program Status

- **Current:** In production; fieldings began in May 2005

Recent and Projected Activities

- **FY07:** Continue fielding
- **FY08:** Continue fielding

ACQUISITION PHASE

Concept & Technology Development | System Development & Demonstration | **Production & Deployment** | Operations & Support

Lightweight Water Purification (LWP) System

FOREIGN MILITARY SALES
None

CONTRACTORS
Mechanical Equipment Co. (MECO) (New Orleans, LA)

Line Haul Tractor

Supports corps and division rear supply activities with transportation of bulk petroleum products, containerized cargo, general cargo, and bulk water.

INVESTMENT COMPONENT

Modernization

Recapitalization

Maintenance

Description & Specifications

The M915A3 Line Haul Tractor is the Army's key line haul distribution platform. It is a 6 x 4 tractor with a 2-inch kingpin and 105,000-pound gross combination weight capacity. The vehicle is transportable by highway, rail, marine, and air modes worldwide. It features the following:

- 52,000 pounds gross vehicle weight/ 105,000 pounds gross combination weight
- 2-inch, 30,000-pound fifth-wheel capacity
- Electronic diagnosis
- Anti-lock brake system
- Sixty-five mph towing speed with full payload
- Engine, Detroit Diesel S60 (430 hp, 1450 lb-ft torque, DDEC IV engine controller)
- Transmission, Allison HD5460P (6-speed, automatic) with power take off.

The M915A4 truck tractor is an M915 truck tractor rebuilt using the glider kit. It is a non-developmental item (NDI)used primarily to transport the M871, 22 1/2-ton flatbed semi-trailer; the M872, 34-ton flatbed semi-trailer; the M1062A1, 7500 gallon semi-trailer; and the M967/M969, 5000-gallon semi-trailer.

The M916A3 Light Equipment Transport (LET) is a 68,000-pound gross vehicle weight tractor with 3-1/2-inch, 40,000-lb capacity, compensator fifth wheel. It has an electronic diesel engine, automatic electronic transmission, anti-lock brakes, and air-conditioning, and is capable of operating at speeds up to 55 mph. This Non-Developmental Item (NDI) vehicle is used primarily to transport the M870 40-ton low-bed semi-trailer.

Program Status

- **Current:** Both vehicles in production
- **FY06:** Market survey for a next-generation vehicle concept

Recent and Projected Activities

- **FY06–FY08** Production of M915A3 vehicles (contract buyout)
- **FY06–FY07:** Requirements generation and systems development; develop purchase descriptions
- **FY08:** Close out current contract; award contract for M915 family of vehicles re-buy

ACQUISITION PHASE

Concept & Technology Development | System Development & Demonstration | Production & Deployment | Operations & Support

Line Haul Tractor

FOREIGN MILITARY SALES
Afghanistan

CONTRACTORS
Freightliner Trucks (Portland, OR)
Detroit Diesel (Detroit, MI)
Meritor (Troy, MI)
Holland Hitch (Holland, MI)

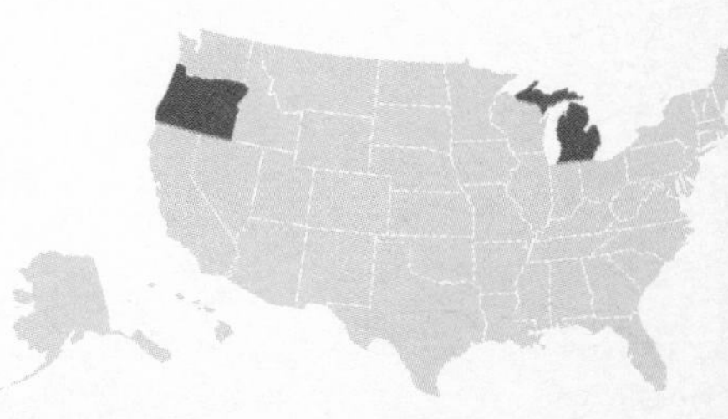

Load Handling System Compatible Water Tank Rack (Hippo)

Provides a mobile, hardwall water tanker to perform bulk distribution of potable water to the division and brigade areas.

INVESTMENT COMPONENT

Modernization

Recapitalization

Maintenance

Description & Specifications

The Load Handling System Compatible Water Tank Rack (Hippo) represents the latest technology in bulk water distribution systems. It replaces the 3K and 5K Semi-trailer Mounted Fabric Tanks (SMFTs). Hippo consists of a 2,000 gallon potable water tank in an ISO frame with an integrated pump, engine, alternator, filling stand, and 70-foot hose reel with bulk suction and discharge hoses. It has the capacity to pump 125 gallons of water per minute.

Hippo is fully functional mounted or dismounted and is transportable when full, partially full, or empty. It is designed to operate in cold weather environments and can prevent water from freezing at -25 degrees Fahrenheit. Hippo can be moved and set up rapidly. In addition, it can be established using minimal assets and personnel.

Program Status

- **4QFY06:** New contract awarded; conditional materiel release

Recent and Projected Activities

- **FY07:** Full materiel release
- **FY07:** First unit equipped
- **FY08:** Continue production and fielding
- **FY09:** Continue production and fielding

ACQUISITION PHASE

Concept & Technology Development | System Development & Demonstration | **Production & Deployment** | Operations & Support

Load Handling System Compatible Water Tank Rack (Hippo)

FOREIGN MILITARY SALES
None

CONTRACTORS
Mil-Mar Century, Inc. (Dayton, OH)
WEW Westerwalder Eisenwerk (Weitefeld, Germany)

Longbow Apache

Conducts close combat attack, deep precision strikes, and armed reconnaissance and security in day, night, and adverse weather conditions.

INVESTMENT COMPONENT

Modernization

Recapitalization

Maintenance

Description & Specifications

The AH-64D Longbow Apache is the Army's heavy attack platform for both the current and Future Force. It is highly mobile, lethal, and can destroy armor, personnel, and materiel targets in obscured battlefield conditions. The fleet includes the A model Apache and D model Longbow. The Longbow remanufacturing effort uses the A model and incorporates a millimeter wave fire control radar (FCR), radar frequency interferometer (RFI), fire-and-forget radar-guided Hellfire missile, and other cockpit management and digitization enhancements.

Both A and D models are undergoing recapitalization modifications such as upgrading to second-generation forward looking infrared (FLIR) technology with the Arrowhead Modernized Target Acquisition Designation Sight/Pilot Night Vision Sensor (MTADS/PNVS), non-line-of-sight communications, and video transmission/reception, and to reduce maintenance cost. Fielding (634 Apaches) began in June 2005.

Apache is fielded to active National Guard (NG) and Army Reserve (AR) attack battalions and cavalry units through the 2006 Army Modernization Plan. The Army will convert its remaining A models to the Longbow Apache configuration. The program began with two multi-year contracts: the first delivered 232 Longbows from FY96–FY01; the second delivered an additional 269 aircraft from FY02–FY06; 120 A to D conversions will occur in FY07–FY10.

The current strategy is to upgrade 597 Block I and II AH-64Ds to a Block III configuration, with an eventual acquisition objective of 634 total Block III modernized Longbows. The Block III modernized Longbows will be designed and equipped with an opensystems architecture to incorporate the latest communications, navigation, sensor, and weapon systems.

Combat mission speed: 167 mph
Combat range: 300 miles
Combat endurance: 2.5 hours
Max. gross weight: 20,260 pounds
Armament: Hellfire missiles, 2.75-inch rockets, and 30mm chain gun
Crew: Two (pilot and co-pilot gunner)

Program Status

- **3QFY06:** Block III Milestone B DAB approval allowing entry in SDD
- **FY06:** Fielded two attack battalions

Recent and Projected Activities

- **FY07:** Field last active component AH-64D attack battalion
- **FY07:** Field First National Guard (AZ) AH-64D attack battalion
- **FY07:** Contract award for 120 extended Block II A to D conversions
- **FY07:** Award production contract for 11 wartime replacement Block II aircraft, bringing new builds to 45

ACQUISITION PHASE

Concept & Technology Development | System Development & Demonstration | Production & Deployment | Operations & Support

Longbow Apache

FOREIGN MILITARY SALES
Egypt, Greece, Israel, Kuwait, Netherlands, Saudi Arabia, Singapore, United Arab Emirates
Direct commercial sales: Japan, Greece, and the United Kingdom

CONTRACTORS
Airframe/fuselage:
Boeing (Mesa, AZ)
Fire Control Radar:
Northrop Grumman (Linthicum, MD)
Lockheed Martin (Oswego, NY; Orlando, FL)
MTADS/PNVS:
Lockheed Martin (Orlando, FL)
Boeing (Mesa, AZ)
Rotor blades:
Ducommun AeroStructures (Monrovia, CA)

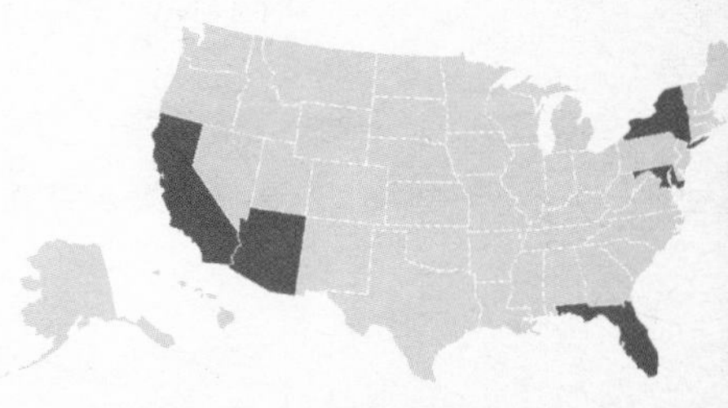

M2 Machine Gun Quick Change Barrel Kit

Provides a quick-change barrel capability that increases Soldier lethality and survivability on the battlefield.

INVESTMENT COMPONENT

Modernization

Recapitalization

Maintenance

Description & Specifications

The M2 Machine Gun Quick Change Barrel Kit will enable the Soldier to quickly change the barrel without the need to reset the headspace and timing. Improper headspace adjustment can cause improper function of the gun and, frequently, damage to parts or injury to personnel. It contains an active safety device and a flash hider to reduce weapon signature. It improves Soldier effectiveness and increases Soldier safety and survivability. It enables faster target engagement because the Soldier needs less time to change the barrel. This improvement increases Soldier safety and survivability.

Program Status

- **Current:** The Quick Change Barrel is undergoing full and open competition

Recent and Projected Activities

- **2QFY07:** Release request for proposal for open competition of the Quick Change Barrel Kit
- **2QFY07:** Conduct initial test and evaluation for down select
- **3QFY07:** Conduct limited testing on two selected vendors
- **4QFY07:** Down select to one vendor; award contract for Quick Change Barrel Kit hardware
- **1QFY08:** Conduct developmental test and evaluation
- **2QFY08:** Conduct user assessment
- **4QFY08:** Milestone C/type classification
- **1QFY09:** Production award

ACQUISITION PHASE

Concept & Technology Development | System Development & Demonstration | Production & Deployment | Operations & Support

M2 Machine Gun Quick Change Barrel Kit

FOREIGN MILITARY SALES
None

CONTRACTORS
To be determined

Maneuver Control System (MCS)

Distributes technical information on the battlefield, allowing commanders to readily access and display current situation reports, intelligence, and contact reports that assess enemy strength and movement, as well as the status of friendly forces.

INVESTMENT COMPONENT

- Modernization
- Recapitalization
- Maintenance

Description & Specifications

The Maneuver Control System (MCS) is the commander and staff's primary tactical planning tool from battalion through corps. MCS provides automated support for planning, coordinating, controlling, and using maneuver functional area assets and tasks.

MCS uses a suite of products to provide a Common Operational Picture (COP) to aid commanders and staffs in their decision making. These products include digital maps, aerial and satellite photos, 3-D flyover views of the battlespace, mobility analysis of the terrain, and map overlays with intelligence and battle resources by unit. Commanders can quickly analyze different courses of action and make decisions based on up-to-date situational assessments, maneuver schemes, doctrine, and changes encountered during battle.

When a plan is determined, MCS prepares and sends warnings, operations orders, and related annexes. During the mission, MCS provides automatic updates of friendly/enemy unit movement locations. Because all information is maintained on the system's database, re-tasking of units is rapid and flexible. As an integrated part of the battle command systems, MCS receives information from external sources and updates them within predetermined operational plans. Such sources include intelligence, fire support, supply status, and air operations requests. To simplify its operation, the system uses commericial office applications to generate reports and display images, charts, and graphics.

Advanced configuation MCS (Block III and IV) supports Command Post of the Future (CPOF), which allows users to share their workspace map displays and data between all units equipped with this capability. It includes interactive whiteboard planning and war games simulation.

Program Status

- **3QFY06:** CPOF transitioned from DARPA to Project Manager Battle Command
- **4QFY06:** MCS capability production document updated to include full suite of tactical battle command requirements and completes Army staffing
- **1QFY07:** CPOF begins program transition to Tactical Battle Command (TBC) as a technical insertion to MCS
- **1QFY07:** CPOF obtains favorable Milestone B decision for development of a full spectrum collaboration capability and receives favorable Milestone C decision to begin Army-wide fielding of the current product
- **1QFY07:** JROC approves MCS 6.4 CPD

Recent and Projected Activities

- **2QFY07:** CPOF begins Army-wide fielding

ACQUISITION PHASE

Concept & Technology Development | System Development & Demonstration | **Production & Deployment** | Operations & Support

Maneuver Control System (MCS)

FOREIGN MILITARY SALES
None

CONTRACTORS
Software development:
Lockheed Martin (Tinton Falls, NJ)
General Dynamics
CECOM Software Engineering Center
(Fort Monmouth, NJ)
Viecore (Tinton Falls, NJ)
Northrop Grumman (San Diego, CA)

Medical Communications for Combat Casualty Care (MC4)

Integrates, fields, and supports a medical information management system for Army tactical medical forces, enhancing medical situational awareness for operational commanders.

INVESTMENT COMPONENT

Modernization

Recapitalization

Maintenance

Description & Specifications

The Medical Communications for Combat Casualty Care (MC4) system is a joint theater-level, automated combat health support system for the tactical medical forces. It serves three distinct user communities: warfighter commanders, healthcare providers, and medical staffs in theater. The system enhances medical situational awareness for the operational commander, enabling a comprehensive, life-long electronic medical record for all Service members. Using the Theater Medical Information Program (TMIP)-joint software, MC4 receives, stores, processes, transmits, and reports medical command and control, medical surveillance, casualty movement/tracking, medical treatment, medical situational awareness, and medical logistics data across all levels of care.

The MC4 system provides the Army's solution to the Title 10 requirement for a medical tracking system for all deployed Service members. The MC4 system is a fully operational standard Army system that operates on commercial off-the-shelf hardware. It supports commanders with a streamlined personnel deployment system using digital medical information.

The MC4 system is composed of seven Army-approved line items that can be configured to support levels 1-4 of the health care continuum. Future MC4 enhancements will be accomplished through system upgrades and Pre-Planned Product Improvements (P3I). The MC4 program completed a successful a full rate production decision review on July 21, 2005.

Program Status

- **3QFY06:** Currently fielding per Army Resourcing Priority List

Recent and Projected Activities

- **1QFY08:** First P3I
- **3QFY08:** Second P3I

ACQUISITION PHASE

Concept & Technology Development | System Development & Demonstration | Production & Deployment | Operations & Support

Medical Communications for Combat Casualty Care (MC4)

FOREIGN MILITARY SALES
None

CONTRACTORS
Hardware:
GTSI (Chantilly, VA)
CDW-G (Chicago, IL)
System engineering support:
Johns Hopkins University Applied Physics Laboratory (Laurel, MD)
System integration support:
L-3 Communications Titan Group (Reston, VA)
Fielding, training, and system administration support:
General Dynamics (Fairfax, VA)

Medium Caliber Ammunition

Provides warfighters with overwhelming lethality overmatch in medium caliber ammunition for Current and Future Force systems.

INVESTMENT COMPONENT

- **Modernization**
- Recapitalization
- Maintenance

Description & Specifications

Medium caliber ammunition includes 20mm, 25mm, 30mm, and 40mm armor-piercing, high-explosive, smoke, illumination, CS (tear gas), practice, and antipersonnel cartridges with the capability to defeat light armor, materiel, and personnel targets. Target practice cartridges are used to train Soldiers in the use of warfighting rounds. The 20mm cartridge is used in the Counter Rocket, Artillery, and Mortar (CRAM) weapon system. The 25mm cartridges are fired from the M242 Bushmaster gun on the Bradley Fighting Vehicle. The 30mm cartridges are used in the Apache helicopter's M230 Chain Gun. A variety of 40mm cartridges are designed for use in the MK19 Grenade Machine Gun and the M203 Grenade Launcher.

Program Status

- **Current:** In production

Recent and Projected Activities

- **2QFY07:** Multiple year family buy for all 25mm and 30mm training ammunition

ACQUISITION PHASE

Concept & Technology Development | System Development & Demonstration | **Production & Deployment** | Operations & Support

Medium Caliber Ammunition

FOREIGN MILITARY SALES

20mm: Australia, Bahrain, Finland, Israel, Jordan, Oman, Singapore

30mm: United Arab Emirates

40mm: Australia, Colombia, Israel, Jordan

CONTRACTORS

General Dynamics Ordnance and Tactical Systems (Marion, IL; Red Lion, PA)

Alliant Techsystems (Radford, VA; Rocket City, WV)

American Ordnance (Milan, TN)

AMTEC Corp. (Janesville, WI)

DSE (Balimoy) Corp. (Tampa, FL)

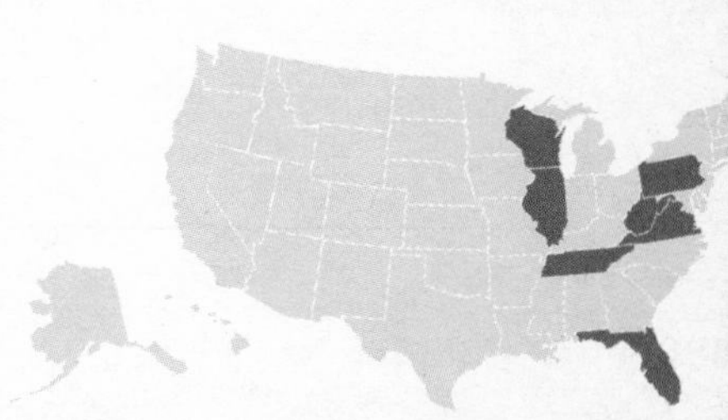

Medium Extended Air Defense System (MEADS)

Provides low- to medium-altitude air and missile defense to maneuver forces and other land component commanders' designated critical assets for all phases of tactical operations.

INVESTMENT COMPONENT

- Modernization
- Recapitalization
- Maintenance

Description & Specifications

The Medium Extended Air Defense System (MEADS) provides a robust, 360-degree defense using the PAC-3 hit-to-kill Missile Segment Enhancement (MSE) against the full spectrum of theater ballistic missiles, anti-radiation missiles, cruise missiles, unmanned aerial vehicles, tactical air-to-surface missiles, and rotary and fixed wing threats. MEADS will also provide:

- Defense against multiple and simultaneous attacks by short-range ballistic missiles, low-radar cross-section cruise missiles, and other air-breathing threats
- Immediate C-130 and C-17 deployment for early entry operations, and lift capability by CH-47 helicopters and Marine Corps Landing Craft Air Cushion and Landing Craft Utility
- Mobility to displace rapidly and protect maneuver forces assets during offensive operations
- Netted, distributed, and open architecture and modular components to increase survivability and flexibility of use in a number of operational configurations
- A significant increase in firepower with the PAC-3 MSE, with greatly reduced requirements for manpower, maintenance, and logistics

The MEADS weapon system will use its netted and distributed architecture to ensure joint and allied interoperability, and to enable a seamless interface to the next generation of battle management command, control, communications, computers, and intelligence (BMC4I). The system's improved sensor components and its ability to link other airborne and ground-based sensors facilitate the employment of its battle elements.

The MEADS weapon system's objective battle management tactical operations center (TOC) will provide the basis for the future common air and missile defense (AMD) TOC, leveraging modular battle elements and a distributed and open architecture to facilitate continuous exchange of information to support a more effective AMD system of systems.

Program Status

- **1QFY06:** Integrated baseline review
- **1QFY07:** Launcher system requirement review
- **1QFY07:** Launcher system requirement review

Recent and Projected Activities

- **2QFY07–3QFY07:** Major end items preliminary design review (PDR)
- **4QFY07:** System PDR
- **1QFY10:** System critical design review

ACQUISITION PHASE

- Concept & Technology Development
- System Development & Demonstration
- Production & Deployment
- Operations & Support

Medium Extended Air Defense System (MEADS)

FOREIGN MILITARY SALES
None

CONTRACTORS
MEADS International is a multinational joint venture headquartered in Orlando, FL. MEADS International's participating companies are MBDA-Italia, the European Aeronautic Defence and Space Company (EADS) and Lenkflugkorpersysteme (LFK) in Germany, and Lockheed Martin in the United States

Meteorological Measuring Set-Profiler (MMS-P)

ncreases the lethality of all field artillery platforms by providing modern, real-time meteorological data over an xtended battlespace.

NVESTMENT COMPONENT

Modernization

Recapitalization

Maintenance

Description & Specifications

The AN/TMQ-52 Meteorological Measuring Set–Profiler (MMS-P) uses a suite of meteorological (MET) sensors and MET data from communications satellites, along with an advanced weather model, to provide highly accurate MET data for a wide range of deep fire weapons and munitions.

Profiler measures and transmits MET conditions, such as wind speed, wind direction, temperature, pressure and humidity, rate of precipitation, visibility, cloud height, and cloud ceiling, that are required for precise targeting and terminal guidance. Profiler uses this information to build a four-dimensional MET model (height, width, depth, and time) that includes terrain effects.

By providing accurate MET messages, Profiler gives the artillery a greater probability of first-round hit with indirect fire. This new capability increases the lethality of field artillery platforms such as the Multiple Launch Rocket System (MLRS), Paladin, and self-propelled or towed howitzers, and produces significant savings for the Army.

The system is housed in a Standard Integrated Command Post System (SICPS) rigid wall shelter and transported on an M1152 A1 High Mobility Multipurpose Wheeled Vehicle (HMMWV). The system uses common hardware, software, and operating systems. The initial configuration provides MET data throughout a 60-kilometer radius, while the follow-on variant extends coverage to 500 kilometers. For the first time, the artillery community has the capability of applying MET data along the trajectory from the firing platform to the target area.

Program Status

- **FY06:** Full rate production and fielding to Modular Brigade Combat Teams (MBCTs)
- **FY07:** Full rate production and fielding to MBCTs

Recent and Projected Activities

- **FY08:** Continue full rate production and fielding to MBCTs

ACQUISITION PHASE

Concept & Technology Development | System Development & Demonstration | Production & Deployment | Operations & Support

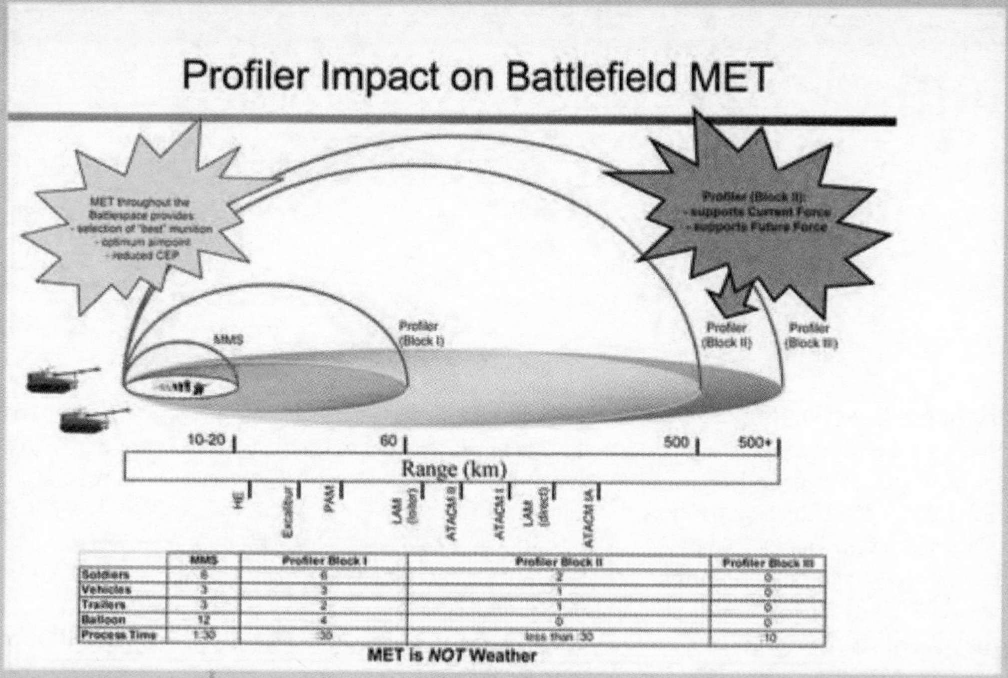

Meteorological Measuring Set-Profiler (MMS-P)

Profiler Impact on Battlefield MET

MET throughout the Battlespace provides:
- selection of "best" munition
- optimum aimpoint
- reduced CEP

Profiler (Block II):
- supports Current Force
- supports Future Force

MMS
Profiler (Block I)
Profiler (Block II)
Profiler (Block III)

10-20 | 60 | 500 | 500+

Range (km)

HE, Excalibur, PAM, LAM (loiter), ATACM II, ATACM I, LAM (direct), ATACM IA

	MMS	Profiler Block I	Profiler Block II	Profiler Block III
Soldiers	6	6	2	0
Vehicles	3	3	1	0
Trailers	3	2	1	0
Balloon	12	4	0	0
Process Time	1:30	:30	less than :30	:10

MET is *NOT* Weather

FOREIGN MILITARY SALES

None

CONTRACTORS

Smiths Detection (Edgewood, MD)

Penn State University (University Park, PA)

Modular Fuel System (MFS)

Provides the capability to receive, store, transport, distribute, issue, surge, and redistribute fuel between brigades, refuel on the move operations, and deploy without construction support.

INVESTMENT COMPONENT

Modernization

Recapitalization

Maintenance

Description & Specifications

The Modular Fuel System (MFS), formerly known as the Load Handling System Modular Fuel Farm (LMFF), is transported by the Heavy Expanded Mobile Tactical Truck Load Handling System (HEMTT-LHS) and the Palletized Loading System. It is composed of 14 each 2,500-gallon capacity tank rack modules and two each pump/filtration modules. Each tank rack module has a baffled 2,500-gallon fuel storage tank and onboard storage compartments for hoses, nozzles, fire extinguishers, and grounding rods. The pump filtration module includes a 600 gallons per minute (GPM) diesel engine-driven centrifugal pump, filter separator, valves, fittings, hoses, refueling nozzles, and a manual hand pump for gravity discharge operations. Each pump filtration module has onboard storage for hoses, ground rods, water cans and fire extinguishers. The pump module has an evacuation capability that allows the hoses in the system to be purged of fuel prior to recovery. The MFS's configuration can vary in size (total capacity) based on the type of force supported.

Program Status

- **3QFY05:** Milestone C low rate initial production contract awarded
- **1QFY06:** Production qualification testing
- **3QFY06:** Limited user testing

Recent and Projected Activities

- **FY07:** Full rate production decision
- **3QFY07:** First unit equipped
- **FY08:** Production option award
- **FY09:** Production option award

ACQUISITION PHASE

Concept & Technology Development | System Development & Demonstration | Production & Deployment | Operations & Support

Modular Fuel System (MFS)

FOREIGN MILITARY SALES
None

CONTRACTORS
E.D. Etnyre and Co. (IL)
DRS Sustainment Systems, Inc. (St. Louis, MO)

Mortar Systems

Enhances mission effectiveness of the maneuver unit commander by providing organic indirect fire support.

INVESTMENT COMPONENT

Modernization

Recapitalization

Maintenance

Description & Specifications

The U.S. Army uses three variants of 120mm mortar systems. All are smoothbore, muzzle-loaded weapons in mounted/dismounted configurations. The M120 120mm Towed Mortar System (TMS) mounts on the M1101 Trailer and is emplaced/displaced using a "Quick Stow" system. The mounted variants are the M121 120mm Mortar and the 120mm Recoiling Mortar System.

The **M252 81mm Mortar System features** a high rate of fire, extended range, and improved characteristics.

The **M224 60mm Mortar System** is a lightweight, man-portable, mortar with improved rate-of-fire capabilities. It can be drop-fired from the standard baseplate or handheld and trigger-fired.

The **M95/M96 Mortar Fire Control System (MFCS)** combines a fire control computer with an inertial navigation and pointing system, allowing crews to fire in under a minute, greatly improving mortar lethality.

The **Lightweight 81mm Mortar Program's** goal is to develop a mortar system 30 to 40 percent lighter than the current M252 81mm system.

The **M32 Lightweight Handheld Mortar Ballistic Computer (LHMBC)** has a tactical modem and embedded GPS so mortar crews can send and receive digital calls for fire messages, calculate ballistic solutions, and navigate. It can be fielded to 60mm, 81mm, and 120mm towed units as an M23 replacement.

Program Status

- **3QFY06:** Initial production of 134 120mm mortar systems completed
- **3QFY06–1QFY07:** 120mm mortar systems fielded to 12+ Infantry Brigade Combat Teams (BCTs)
- **3QFY06–1QFY07:** MFCS fielded to four Heavy BCTs and two Stryker BCTs
- **3QFY06–1QFY07:** LHMBC fielded to nine Infantry BCTs

Recent and Projected Activities

- **3QFY07:** Initiate fielding of 120mm Mortar Quick-Stow Device
- **2QFY07–4QFY08:** Complete production and fielding of M95/M96 MFCS and M32 LHMBC
- **2QFY07–4QFY08:** Complete production and fielding of 60mm and 120mm weapon
- **2QFY09:** Initiate fielding of Dismounted 120mm MFCS

ACQUISITION PHASE

Concept & Technology Development | System Development & Demonstration | **Production & Deployment** | Operations & Support

Commander's Interface
SINCGARS Radio
Power Distribution Assembly
Gunner's Display
MFCS V1 Software
Driver's Display
Pointing Devices

Lightweight Handheld Mortar Ballistic Computer (LHMBC)

Operation Enduring Freedom

Mortar Modularity

M120 120mm Mortar w/M1101 Trailer and Stowage Device

M224 60mm Mortar

M67 Sight

M1064A3 Mortar Carrier

SBCT Mortar Carrier

FCS NLOS Mortar Carrier

Mortar	Range (meters)	Weight (pounds)	Rate of Fire (rounds per minute)	Crew	Ammunition
120mm	7240	319	16 for the first minute 4 sustained	4 - M121 carrier-mounted 5 - M120 towed	High explosive (HE) (M934A1), White Phosphorous (WP) smoke (M929), illumination (visible light, M930 and infrared [IR], M983), and full-range practice (FRP) (M931)
M252 81mm	5935	90	30 first two minutes 15 sustained	3	HE (M821A2), Red Phosphorous smoke (M819), illumination (visible light, M853A1 and IR, M816), and FRP (M879)
M224 60mm	3489	46.5 (conventional), 18.0 (handheld)	30 first four minutes 20 sustained	3	HE (M720A1), WP smoke (M722A1), illumination (visible light, M721 and IR, M767), and FRP (M769)

Mortar Systems

FOREIGN MILITARY SALES
None

CONTRACTORS
Mortar weapon system production, staging, and fielding:
General Dynamics Ordnance and Tactical Systems (St. Petersburg, FL)
M95/M96 MFCS hardware integration:
Honeywell Aerospace Electronics (Albuquerque, NM)
M32 LHMBC (R-PDA):
(Tallahassee, FL)

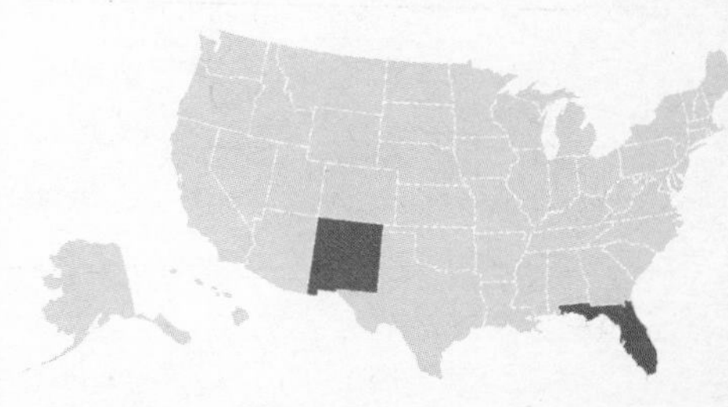

Mounted Warrior

Improves the tactical awareness, lethality, and survivability of combat vehicle crewmen.

INVESTMENT COMPONENT

Modernization

Recapitalization

Maintenance

Description & Specifications

Mounted Warrior (MW) is an integrated system of systems designed for armored vehicle crewmen. MW combines cordless communications and personal displays with Soldier mission equipment, and outfits all crew members (including vehicle commanders, drivers, and gunners) who operate ground platforms. The system leverages capabilities developed in other Warrior programs such as Land Warrior (LW) and Air Warrior (AW). MW-equipped crew members can communicate wirelessly with dismounted LW-equipped Soldiers. The system interfaces with other Army communications and command and control systems. MW includes lightweight, integrated, modular, mission-tailorable equipment and command, control, computers, and communications (C4) devices worn, carried, or used by crewmen when conducting tactical operations with their assigned combat vehicles.

Team Soldier equipped the 4th Battalion, 9th Infantry, 4th Stryker Brigade Combat Team, 2nd Infantry Division at Ft. Lewis, WA, with LW and Mounted Warrior (MW) to conduct a comprehensive assessment. It covered the areas of doctrine, organizations, training, materiel, leadership; education, personnel, and facilities; and tactics, techniques, and procedures. Results of the assessment will be used to determine basis of issue and future Soldier system development decisions. As a result of positive preliminary results, the assessment unit leadership decided to keep the LW and MW systems for training and eventual deployment to combat.

Program Status

- **Current:** Funding decision pending

Recent and Projected Activities

- To be determined

ACQUISITION PHASE

Concept & Technology Development | System Development & Demonstration | Production & Deployment | Operations & Support

Mounted Warrior

FOREIGN MILITARY SALES
None

CONTRACTORS
General Dynamics (Ft. Lauderdale, FL))

Movement Tracking System (MTS)

Provides the Logistics Command with two-way text messaging and position reporting to track and communicate with its mobile assets in real time.

INVESTMENT COMPONENT

Modernization

Recapitalization

Maintenance

MTS on HEMMT

Description & Specifications

The Movement Tracking System (MTS) is a two-way, mobile, wireless satellite system for tacking vehicles and communicating on and off the road during war or peacetime. Messages are transmitted via commercial satellites in near real-time, and vehicle locations are displayed on computers with National Geospatial-Intelligence Agency (NGA) maps. All messages are encrypted end-to-end, including sender and recipient addresses. MTS operates over a variety of geostationary satellites and is designed to transition automatically from one system to another, as required.

MTS comes in two configurations, the V2 mobile unit for vehicle mounting, and the control station configuration for command center operations. Both use the Comtech MT-20 transceiver for sending and receiving messages and position reports. The MT-2011 is a compact, all solid-state device with no moving parts with proven durability. A ruggedized, compact computer is delivered with the vehicle-mount configuration; the control stations use laptops with larger screens for observing mobile units more effectively.

MTS will provide road vehicles as well as watercraft visibility wherever they are deployed. All common user logistic transport (CULT) vehicles, selected combat support (CS), and combat service support (CSS) tactical wheeled vehicles, and watercraft will be fitted with MTS mobile units. A portable MTS unit may be made available to a host nation or foreign national force contributing to a combined operation.

Program Status

- **3QFY06:** Support deployed units in Middle Eastern Theater of Operations; fielding of selected National Guard (NG) brigades
- **4QFY06:** Completed fielding to NG units supporting Homeland Security (Phase I – Hurricane States)
- **1QFY07:** Completed fielding to NG units supporting Homeland Security (Phase II – Northeastern States); completed fielding of 173nd Airborne Brigade (BDE) and 12th Combat Aviation BDE
- **2QFY07:** Completed fielding to U.S. Forces, Korea, and 82nd Airborne

Recent and Projected Activities

- **3QFY07:** Perform form fit checks on LCU 2000 landing craft, LT-128 tug boats, and logistics support vessels (LSV) for TACOM; perform product integration test of MESH technology at MTS test lab
- **4QFY07:** Begin MTS fielding to watercraft vessels in Kuwait and Europe
- **1QFY08:** Continue software development for MTS-ES
- **2QFY08:** Begin testing MTS-ES software
- **3QFY08:** Continue testing MTS-ES software
- **4QFY08:** Field MTS-ES software

ACQUISITION PHASE

Concept & Technology Development | System Development & Demonstration | Production & Deployment | Operations & Support

Movement Tracking System (MTS)

FOREIGN MILITARY SALES
None

CONTRACTORS
System integrator:
COMTECH Mobile Datacom (Germantown, MD)
Software development:
Northrop Grumman (Redondo Beach, CA)
Force XXI Battle Command Brigade and Below (FBCB2) (Ft. Monmouth, NJ)

Night Vision Devices

Enhances the warfighter's visual ability and situational awareness to successfully engage and execute operations day or night, in adverse weather or battlefield obscurant conditions.

INVESTMENT COMPONENT

Modernization

Recapitalization

Maintenance

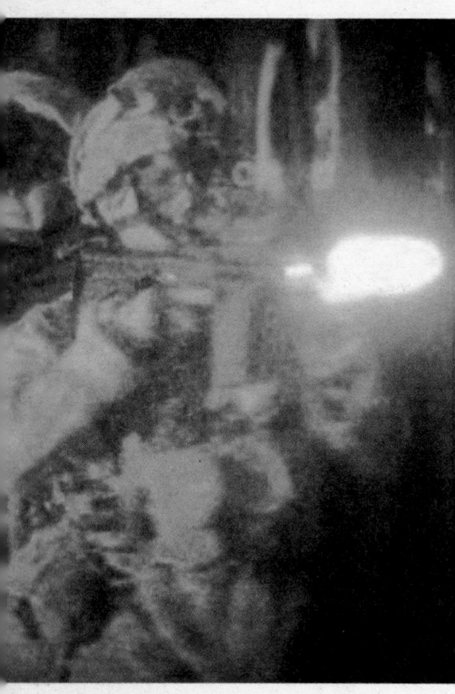

Description & Specifications

The AN/PVS-14 Monocular Night Vision Device is a lightweight, multi-purpose, passive device used by the individual warfighter in close combat, combat support, and combat service support. It amplifies ambient light and very near infrared energy for night operations. AN/PVS-14 can be mounted to the M16/M4 receiver rail.

PVS-14

Field of View: ≥ 40 degrees
Weight (maximum): 0.88 pounds
Magnification: 1x
Range: 150 meters

The Enhanced Night Vision Goggle (ENVG) is a helmet-mounted passive device for the individual Soldier that incorporates image intensification and long-wave infrared sensors into a single, integrated system. ENVG enables missions during daylight, darkness, and degraded battlefield conditions. ENVG improves the Soldier's situational awareness by providing the capability to rapidly detect and recognize man-sized targets while simultaneously maintaining the ability to see detail and to use rifle-mounted aiming lights.

ENVG

Infared sensor detection range: 150 meters
Weight (total system): 2 pounds
Operational time (with one battery set change): 15 hours
Range: 150 meters

Program Status

- **Current:** AN/PVS-14 in production and being fielded
- **FY06:** ENVG optical system in system development and demonstration
- **1QFY07:** ENVG development tests

Recent and Projected Activities

- **Ongoing:** Planned procurement of AN/PVS-14
- **Ongoing:** Fielding in support of Operation Enduring Freedom and Operation Iraqi Freedom
- **Ongoing:** Fielding to support modularity, National Guard, and Reserve
- **2QFY07:** Operational tests

ACQUISITION PHASE

Concept & Technology Development | System Development & Demonstration | Production & Deployment | Operations & Support

Night Vision Devices

FOREIGN MILITARY SALES
Multiple foreign purchasers of AN/PVS-14

CONTRACTORS
AN/PVS-14:
Northrop Grumman (Tempe, AZ; Garland, TX)
ITT Industries (Roanoke, VA)
ENVG:
ITT Industries (Roanoke, VA)

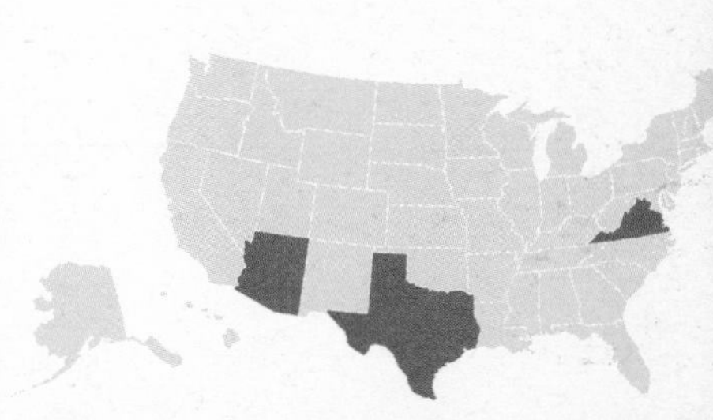

Non-Line of Sight-Launch System (NLOS-LS)

Enhances combat effectiveness and survivability by providing precise, highly deployable, non-line-of-sight lethal fires for the Future Combat Systems (FCS) Brigade Combat Team (BCT).

INVESTMENT COMPONENT

- **Modernization**
- Recapitalization
- Maintenance

Description & Specifications

The Non-Line of Sight-Launch System (NLOS-LS), a core system within the Future Combat Systems (FCS) Brigade Combat Team (BCT) family of systems, provides unmatched lethality and "leap ahead" missile capability for U.S. forces. NLOS-LS consists of precision guided missiles loaded onto a highly deployable, platform-independent container launch unit (CLU) with self-contained technical fire control, electronics, and software to enable remote and unmanned fire support operations.

The precision guided munition being developed is the Precision Attack Missile (PAM). The NLOS-LS CLU will contain 15 missiles. The PAM, which launches vertically from the CLU, will be used primarily to defeat hard and soft, moving, or stationary target elements when fire mission orders are received by current force Advanced Field Artillery Tactical Data System (AFATDS) and the FCS BCT network in future spin-outs. It will be able to receive in-flight target updates via its onboard network radio, and will have limited automatic target recognition capability. PAM will have a multi-functional warhead to effectively engage hard (armor) and soft targets. NLOS-LS will be available for evaluation in FY08 for integration into current forces as part of the FCS BCT spin-out strategy. NLOS-LS also supports the Navy's Littoral Combat Ship against small boat threats. Future missile variants in follow-on FCS BCT spin increments may include air defense and non-lethal capabilities. Key NLOS-LS advantages include the following:

- **Remote** fire control
- **Remote** emplacement
- **Extended-range** target engagements and battle damage assessment
- **Jam-resistant** Global Positioning System
- **Ability** to engage moving targets

Weight: CLU with 15 missiles, approximately 3150 pounds

Dimensions:
Width: 45 inches
Length: 45 inches
Height: 69 inches
Range: Approximately 40 km

Program Status

- **1QFY06:** Preliminary design review
- **3QFY06:** NLOS-LS Navy contract awarded
- **1QFY07:** Critical design review

Recent and Projected Activities

- **2QFY07:** Begin component qualification test
- **1QFY08:** Begin developmental flight test program
- **1QFY08:** Brigade Combat Team evaluations

ACQUISITION PHASE

- Concept & Technology Development
- **System Development & Demonstration**
- Production & Deployment
- Operations & Support

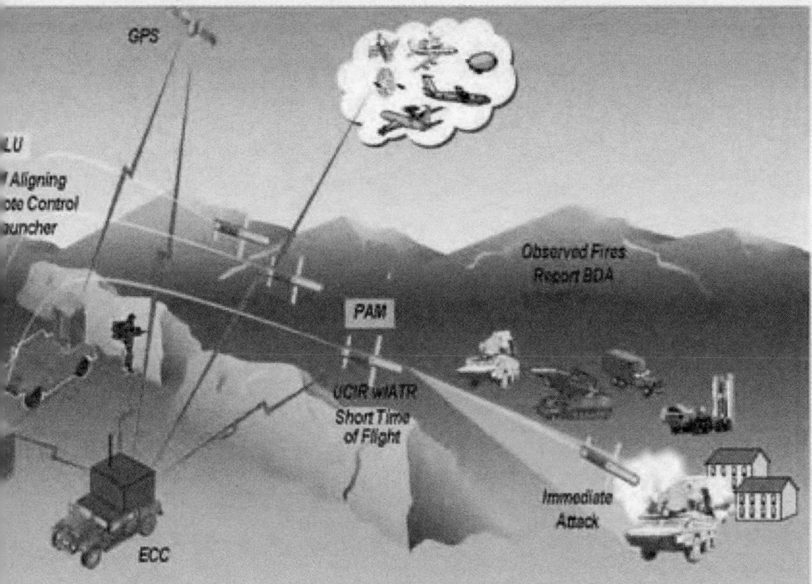

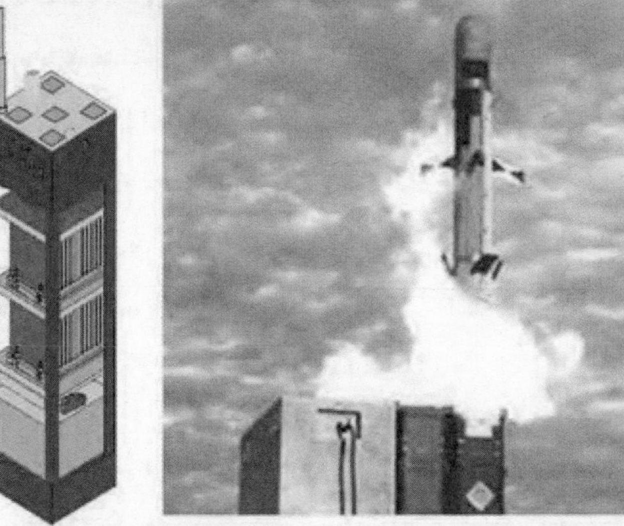

Missile Computer & Commo System

Non-Line of Sight-Launch System (NLOS-LS)

FOREIGN MILITARY SALES
None

CONTRACTORS
Raytheon (Tucson, AZ)
Lockheed Martin Dallas (Dallas, TX)
Lockheed Martin Baltimore (Baltimore, MD)
Aerojet (Gainesville, VA)
L3/IAC (Anaheim, CA)

Precision Attack Munition (PAM)

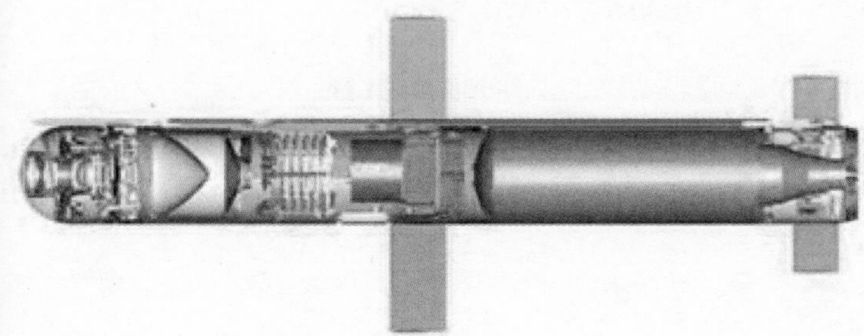

Container/Launch Unit

PM UA Networks / LSI Efforts

- Tactical Fire Control
- Network Development and I/F
- Mission Planning

PEO-TM Efforts

- Development of PAM, LAM, and CLU w/ MCCS
- Technical Fire Control
- Network Interfaces for CLU and Missiles
- Platform Interfaces with LSI and PM FCS

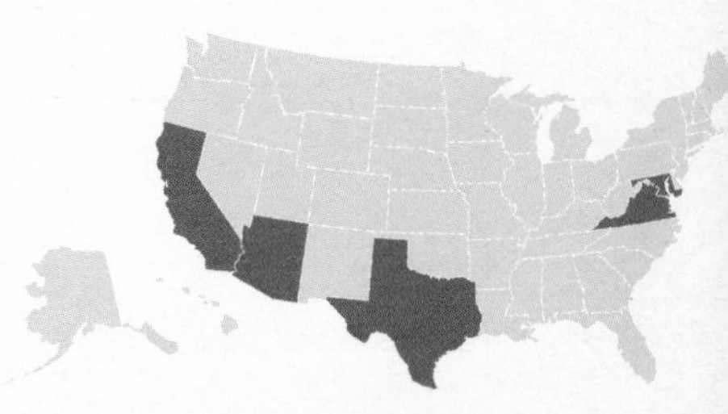

Nuclear Biological Chemical Reconnaissance Vehicle (NBCRV)-Stryker

Delivers accurate, rapid combat information by detecting, sampling, identifying, marking, and reporting the presence of chemical, biological, radiological, and nuclear and toxic industrial material hazards.

INVESTMENT COMPONENT

Modernization

Recapitalization

Maintenance

Description & Specifications

The Nuclear Biological Chemical Reconnaissance Vehicle (NBCRV)-Stryker is the chemical, biological, radiological and nuclear (CBRN) reconnaisance configuration of the infantry carrier vehicle in the Stryker Brigade Combat Team (SBCT).

NCBRV-Stryker uses government off-the-shelf military hardware and some systems still in development. Its sensor suite provides outstanding capability on a common platform by use of a single, integrated reconnaissance and surveillance system. The Joint Service Lightweight Standoff Chemical Agent Detector (JSLSCAD), Chemical Biological Mass Spectrometer (CBMS) Block II, and Force XXI Battle Command Brigade and Below (FBCB) are examples of programs currently in development. Some are in production, such as the Joint Biological Point Detection System, M22 Automatic Chemical Agent Alarm, AN/VDR-2 radiation detection, indication, and computation (RADIAC), AN/UDR 13 pocket RADIAC, and other non-nuclear, biological, and chemical equipment such as the Precision Lightweight Global Positioning System Receiver.

NBCRV will have the capability to detect and collect CBRN and toxic industrial material contamination in its immediate environment, on the move through point detection, and at a distance through the use of a standoff detector. It will automatically integrate contamination information from detectors with input from on-board navigation and meteorological systems and automatically transmit digital nuclear, biological, chemical (NBC) warning messages through the Maneuver Control System to follow-on forces.

NBCRV may replace the need for separate M93A1 Fox NBC reconnaissance systems and biological integrated detection systems.

Program Status

- **1QFY06–4QFY06:** Production verification test
- **4QFY06:** Initial operational test and evaluation

Recent and Projected Activities

- **1QFY07–2QFY07:** Live fire test and evaluation
- **3QFY07:** Full rate production decision

ACQUISITION PHASE

Concept & Technology Development | System Development & Demonstration | Production & Deployment | Operations & Support

Nuclear Biological Chemical Reconnaissance Vehicle (NBCRV)–Stryker

FOREIGN MILITARY SALES
None

CONTRACTORS
Prime vehicle:
General Dynamics Land Systems (Sterling Heights, MI)
Sensor software integrator:
CACI (Manassas, VA)

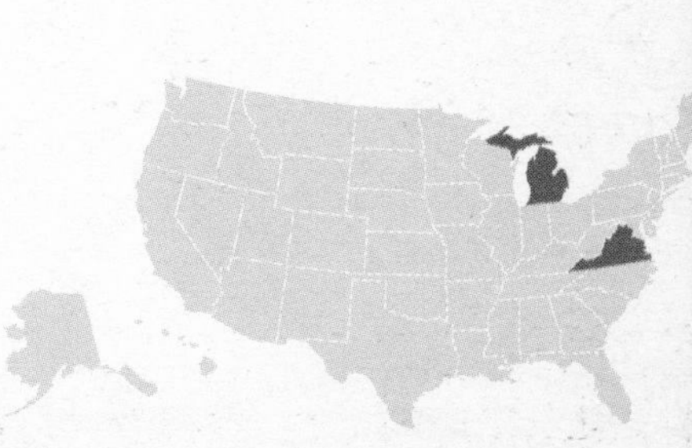

One Semi-Automated Forces (OneSAF) Objective System

Provides simulation software that supports constructive training, mission rehearsal, analysis and research, virtual simulators, and the Future Combat Systems (FCS).

INVESTMENT COMPONENT

Modernization

Recapitalization

Maintenance

Description & Specifications

The One Semi-Automated Forces (OneSAF) Objective System is a computer-generated forces simulation system designed for brigade and below combat and non-combat operations.

OneSAF was built to represent the modular and future force and to represent entities, units, and behaviors across the spectrum of military operations in the Contemporary Operating Environment (COE). OneSAF is unique in its ability to model unit behaviors from fire team to company level for all units.

OneSAF is a cross-domain simulation suitable for supporting training, analyses, research, experimentation, mission planning, and rehearsal activities. Its systems engineering and design for the architecture and software tools provide a universal Army entity-level simulation.

OneSAF will be fully interoperable with the Army's emerging virtual, live, and constructive simulations and will provide next-generation simulation products to support the Army's Future Combat Systems (FCS). It will replace or enhance a variety of existing and legacy simulations currently used by the Army.

Program Status

- **1QFY07:** OneSAF Version 1.0 released

Recent and Projected Activities

- **2QFY07:** Release OneSAF (International) Version 1.0
- **FY08:** Release OneSAF Version 2.0

ACQUISITION PHASE

Concept & Technology Development | System Development & Demonstration | Production & Deployment | Operations & Support

One Semi-Automated Forces (OneSAF) Objective System

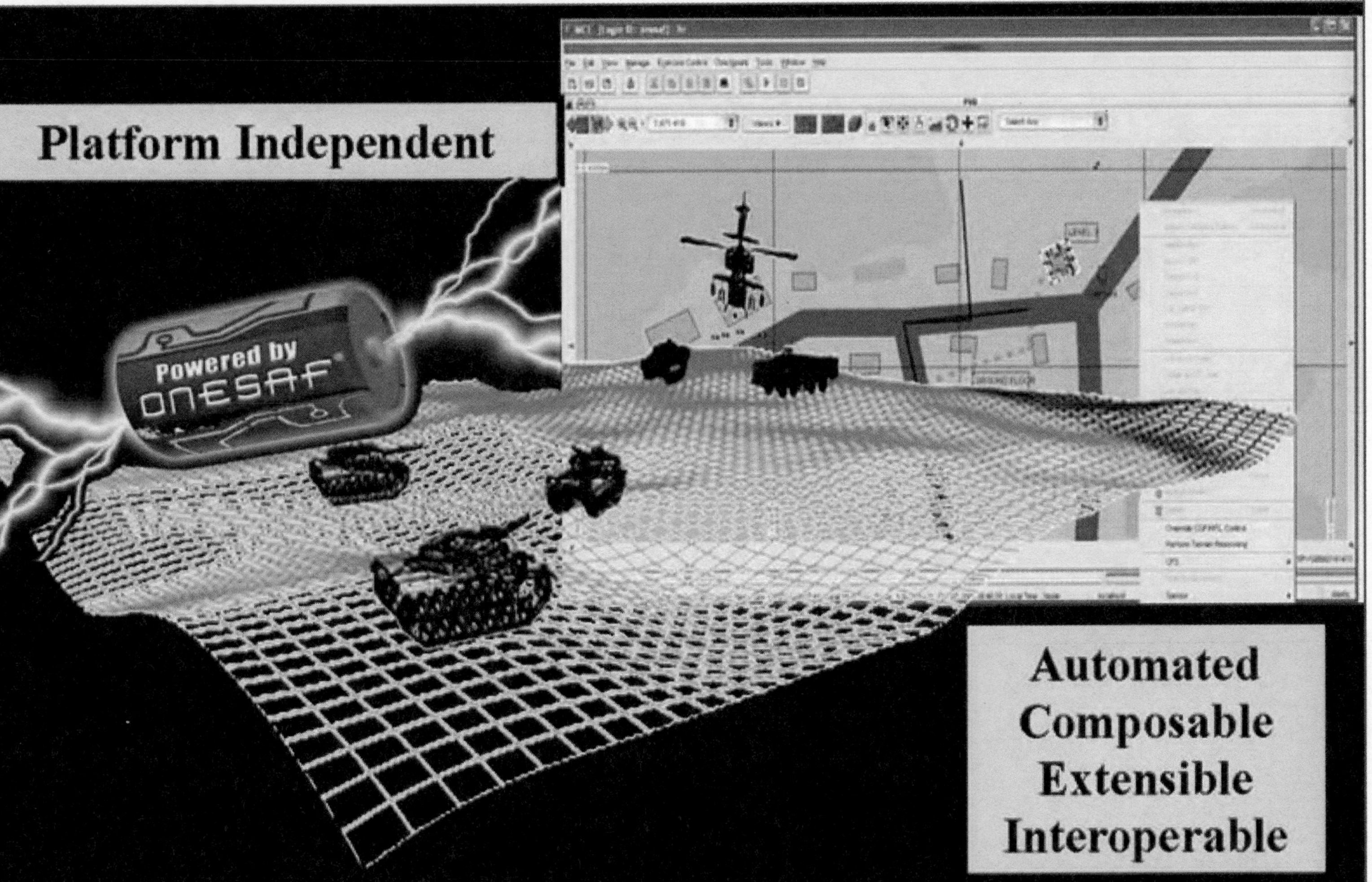

FOREIGN MILITARY SALES
None

CONTRACTORS
SAIC (Orlando, FL)
AcuSoft Inc. (Orlando, FL)
The Aegis Technology Group, Inc. (Orlando, FL)
Northrop Grumman Information Technology (NGIT) (Orlando, FL)
Lockheed Martin (Orlando, FL)

One Tactical Engagement Simulation System (OneTESS)

Supports force-on-force and force-on-target training exercises at brigade and below, for all Battlefield Operating Systems, at installation, Combat Maneuver Training Centers, and deployed sites.

INVESTMENT COMPONENT

- **Modernization**
- Recapitalization
- Maintenance

Description & Specifications

One Tactical Engagement Simulation System (OneTESS) will provide realistic weapons effects of all weapons systems, munitions, countermeasures, and counter-counter measures and stimulate weapon and battlefield sensors while providing the means to objectively assess all weapons effects experienced during live training and testing exercises.

OneTESS is designed to migrate current training from the Multiple Integrated Laser Engagement Simulation (MILES) system to the future technology of geometric pairing, providing a higher fidelity of engagement simulation to the Soldier. The project office will provide a OneTESS standard to facilitate appending/embedding of system capabilities into all existing and future weapons platforms.

OneTESS will provide the following capabilities:

- Ground-to-ground engagements (including directly and indirectly fired munitions, non-line of sight, and beyond-line of sight)
- Electronic warfare and information operations warfare
- Engineer warfare and countermines
- Simulation of nuclear, biological, and chemical engagement and effects
- Ground-to-air engagements
- Air-to-air and air-to-ground engagements, to include Close Air Support
- Smart fire-and-forget engagements
- Directed energy weapons (including high powered microwave and laser)
- Countermeasures and counter-countermeasures
- Non-lethal munitions
- Precision gunnery

Program Status

- **4QFY06:** Preliminary design review

Recent and Projected Activities

- **2QFY07:** Critical design review 1
- **2QFY09:** Limited user test
- **4QFY09:** Milestone C

ACQUISITION PHASE

- Concept & Technology Development
- **System Development & Demonstration**
- Production & Deployment
- Operations & Support

One Tactical Engagement Simulation System (OneTESS)

FOREIGN MILITARY SALES
None

CONTRACTORS
AT&T Government Solutions (Orlando, FL; Vienna, VA)

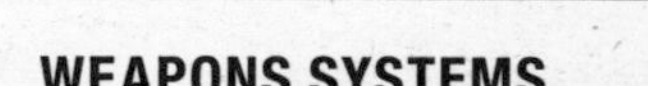

Paladin/Field Artillery Ammunition Supply Vehicle (FAASV)

Provides the primary artillery support for armored divisions, mechanized infantry divisions, and Heavy Brigade Combat Teams.

INVESTMENT COMPONENT

- Modernization
- Recapitalization
- Maintenance

Description & Specifications

The M109A6 (Paladin) 155mm howitzer is the most technologically advanced self-propelled cannon system in the Army. The Field Artillery Ammunition System Vehicle (FAASV) provides an armored ammunition resupply vehicle in support of the Paladin. Paladin uses state-of-the-art components to achieve dramatic improvements in the following:

- **Survivability:** "Shoot and scoot" tactics; improved ballistic and nuclear, biological, and chemical protection
- **Responsive fires:** Capable of firing within 45 seconds from a complete stop with on-board communications, remote travel lock, and automated cannon slew capability
- **Accurate fires:** On-board position navigator and technical fire control
- **Extended range:** 30 kilometers with high-explosive, rocket-assisted projectile and M203 propellant
- **Increased reliability:** Improved engine, track, and diagnostics

Other Paladin specifications include the following:

- **Crew:** Paladin, four; FAASV, five
- **Combat loaded weight:** Paladin, 32 Tons; FAASV, 28 tons
- **Paladin on-board ammo:** 39 rounds
- **FAASV on-board ammo:** 95 rounds
- **Max./sustained rates of fire:** 4/1 rounds/min.
- **Maximum range:** High Explosive Rocket Assisted Projectile (HE/RAP), 22/30 (km)
- **Cruising range:** Paladin, 214 miles; FAASV, 220 miles
- Paladin Digital Fire Control System software supports Fire Support Network

Program Status

- **FY06:** Develop and integrate Excalibur requirements into PDFCS; develop a battlefield digitization trainer
- **FY06:** Installation of Modular Artillery Charge System (MACS) storage racks, PDFCS, and up-powered Auxiliary Power Unit (APU)

Recent and Projected Activities

- **FY07:** Continue Excalibur integration
- **FY07:** Continue installation of MACS storage racks, PDFCS, and APU
- **FY07:** Continue life cycle support for the War on Terrorism (national level reset of 99 Paladins and 86 FAASVs)
- **FY08:** Induct first year's vehicles under Paladin Integrated Management (PIM) program
- **FY08:** Develop detailed strategy plan for PIM program

ACQUISITION PHASE

- Concept & Technology Development
- System Development & Demonstration
- Production & Deployment
- Operations & Support

Paladin/FAASV

FOREIGN MILITARY SALES
None

CONTRACTORS
BAE (York, PA)
Northrop Grumman (Carson, CA)
Anniston Army Depot (Anniston, AL)
Marvin Land Systems (Inglewood, CA)
Kidde Dual Spectrum (Goleta, CA)

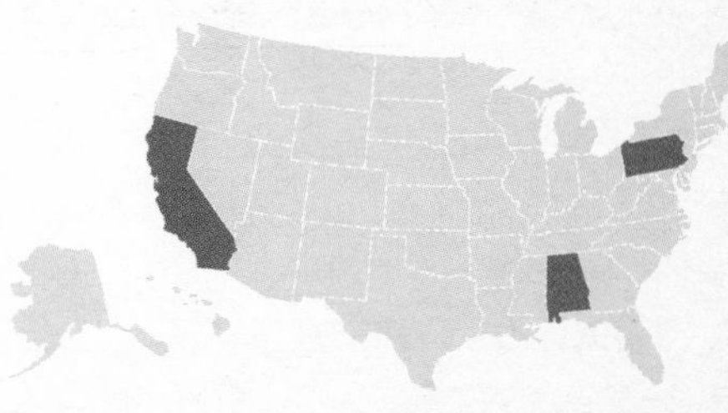

PATRIOT (PAC-3)

Protects ground forces and critical assets at all echelons from advanced aircraft, cruise missiles, and tactical ballistic missiles.

INVESTMENT COMPONENT

Modernization

Recapitalization

Maintenance

Description & Specifications

The PATRIOT (PAC-3) program is an air-defense, guided missile system with long-range, medium- to high-altitude, all weather capabilities designed to counter tactical ballistic missiles (TBMs), cruise missiles, and advanced aircraft. The combat element of the PATRIOT missile system is the fire unit, which consists of a phased array radar set (RS), an engagement control station (ECS), an electric power plant (EPP), an antenna mast group (AMG), a communications relay group (CRG), and eight launching stations (LSs).

The RS provides the tactical functions of airspace surveillance, target detection, identification, classification, tracking, and missile guidance and engagement support. The ECS provides command and control. Each LS contains four ready-to-fire PATRIOT Advanced Capability (PAC-2) guidance enhanced missiles sealed in canisters that serve as shipping containers and launch tubes. PATRIOT's fast-reaction capability, high firepower, ability to track numerous targets simultaneously, and ability to operate in a severe electronic countermeasure environment are significant improvements over previous air defense systems.

The PAC-3 program significantly upgrades the RS and ECS, and adds the new PAC-3 missile. Its primary mission is to kill maneuvering and non-maneuvering TBMs, and counter advanced cruise missile and aircraft threats. The PAC-3 missile uses hit-to-kill technology for greater lethality against TBMs armed with weapons of mass destruction. Up to 16 PAC-3 missiles can be loaded per launcher, increasing firepower and missile defense. The PAC-3 upgrade has improvements that increase performance against evolving threats, meet user requirements, and enhance joint interoperability. The Medium Extended Air Defense System (MEADS) will succeed the PATRIOT system, with the PAC-3 Missile Segment Enhancement (MSE) as the baseline interceptor. PATRIOT and MEADS were combined in FY03 because of shared common components and missions.

Program Status

- **1QFY06:** Post deployment build-6 (PDB-6) development test and evaluation
- **3QFY06:** MSE critical design review
- **3QFY06–2QFY07:** Continue PAC-3 evolutionary development testing
- **1QFY07–09:** Production decision for PAC-3 missile cost reduction initiative configuration
- **1QFY07:** JROC insensitive munitions waiver update

Recent and Projected Activities

- **2QFY07:** PDB-6 initial operational capability
- **3QFY07–4QFY08:** MSE flight testing

ACQUISITION PHASE

Concept & Technology Development | System Development & Demonstration | Production & Deployment | Operations & Support

PATRIOT (PAC-3)

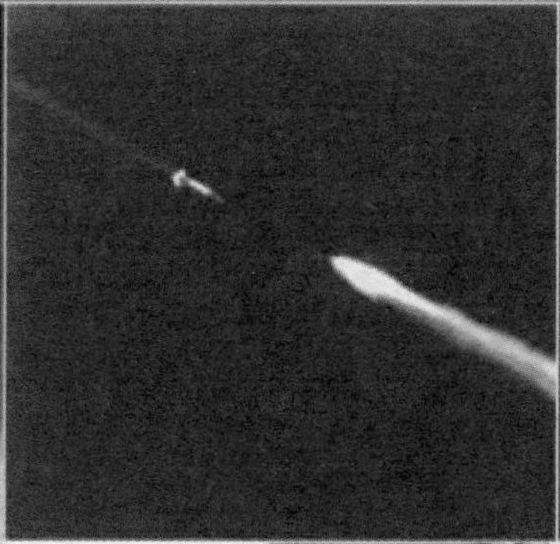

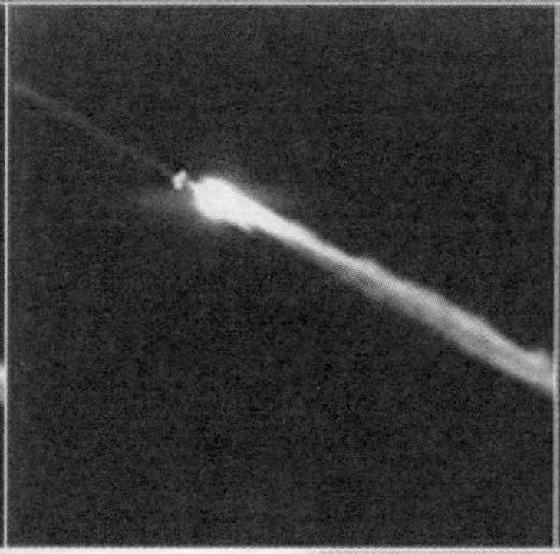

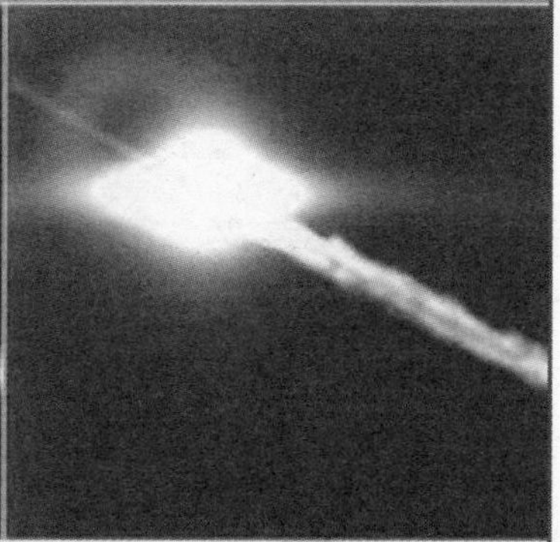

FOREIGN MILITARY SALES
Germany, Greece, Israel, Japan, Kuwait, Saudi Arabia, Spain, Taiwan, and the Netherlands are currently participating in PATRIOT acquisition programs.

CONTRACTORS

PATRIOT system integrator, ground system modifications, recap program:
Raytheon (Andover, MA; Bedford, MA)
PAC-3 Missile, PAC-3 Missile assembly, PAC-3 Missile sub-assembly:
Lockheed Martin (Grand Prairie, TX; Camden, AR; Lufkin, TX)
PAC-3 Missile Seeker:
Boeing (Huntsville, AL)

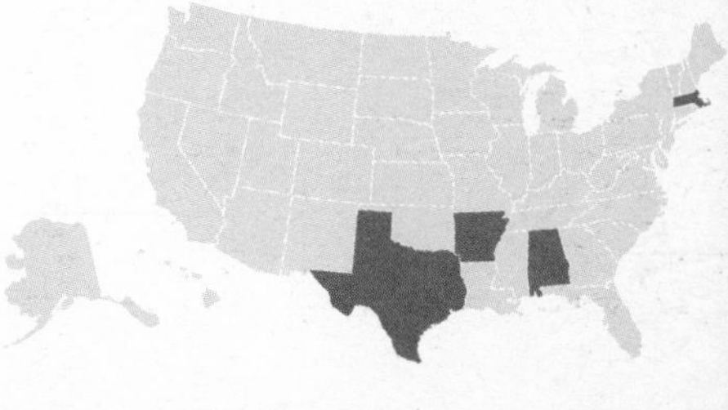

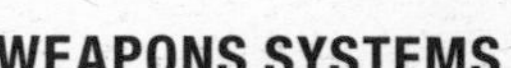

Physical Security Force Protection System of Systems

Detects, assesses, and responds to unauthorized entry or attempted intrusion into installations/facilities or use in tactical environments.

INVESTMENT COMPONENT

Modernization

Recapitalization

Maintenance

Description & Specifications

Integrated Commercial Intrusion Detection System (ICIDS) provides assessment and flexible response to threats. Scalable to accommodate remote intrusion and chemical, biological, radiological, and nuclear (CBRN) agent monitoring. It provides an integrated, secure, intrusion detection system employing commercial interior and exterior sensors, remote area data collectors, close circuit televisions, keypads, biometrics devices, and entry control systems monitored from a primary console.

The Mobile Detection Assessment Response System (MDARS) conducts semi-autonomous random patrols and surveillance to include barrier assessment and theft detection. The MDARS can be used to detect unauthorized access to a facility; verify the status of barriers and products; and investigate alarms from remote locations before dispatching guards.

The Battlefield Anti-Intrusion System (BAIS) consists of a compact, modular, sensor-based warning system that can be used as a tactical stand-alone system. The system consists of a handheld monitor and three seismic/acoustic sensors and provides coverage across a platoon's defensive front (450 meters). It delivers early warning and situational awareness information, classifying detections as personnel, vehicle, wheeled, or tracked intrusions.

The Lighting Kit, Motion Detector (LKMD) is a simple, compact, modular, sensor-based early warning system providing programmable responses of illumination and sound. The LKMD enhances awareness during all types of operations or missions ranging from small-scale contingencies and military operations in urban terrain up to combat.

Program Status

- **FY02–FY06:** ICIDS/BAIS procurement
- **FY06:** LKMD/MDARS developmental testing
- **1QFY07:** fielding of ICIDS/BAIS

Recent and Projected Activities

- **3QFY07:** LKMD/MDARS completion of testing
- **4QFY08:** MDARS/BAIS/ICIDS fieldings

ACQUISITION PHASE

Concept & Technology Development | System Development & Demonstration | Production & Deployment | Operations & Support

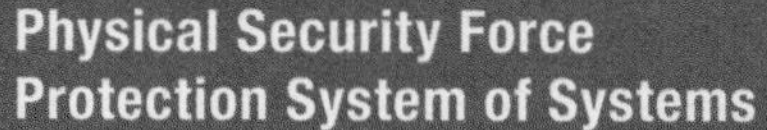

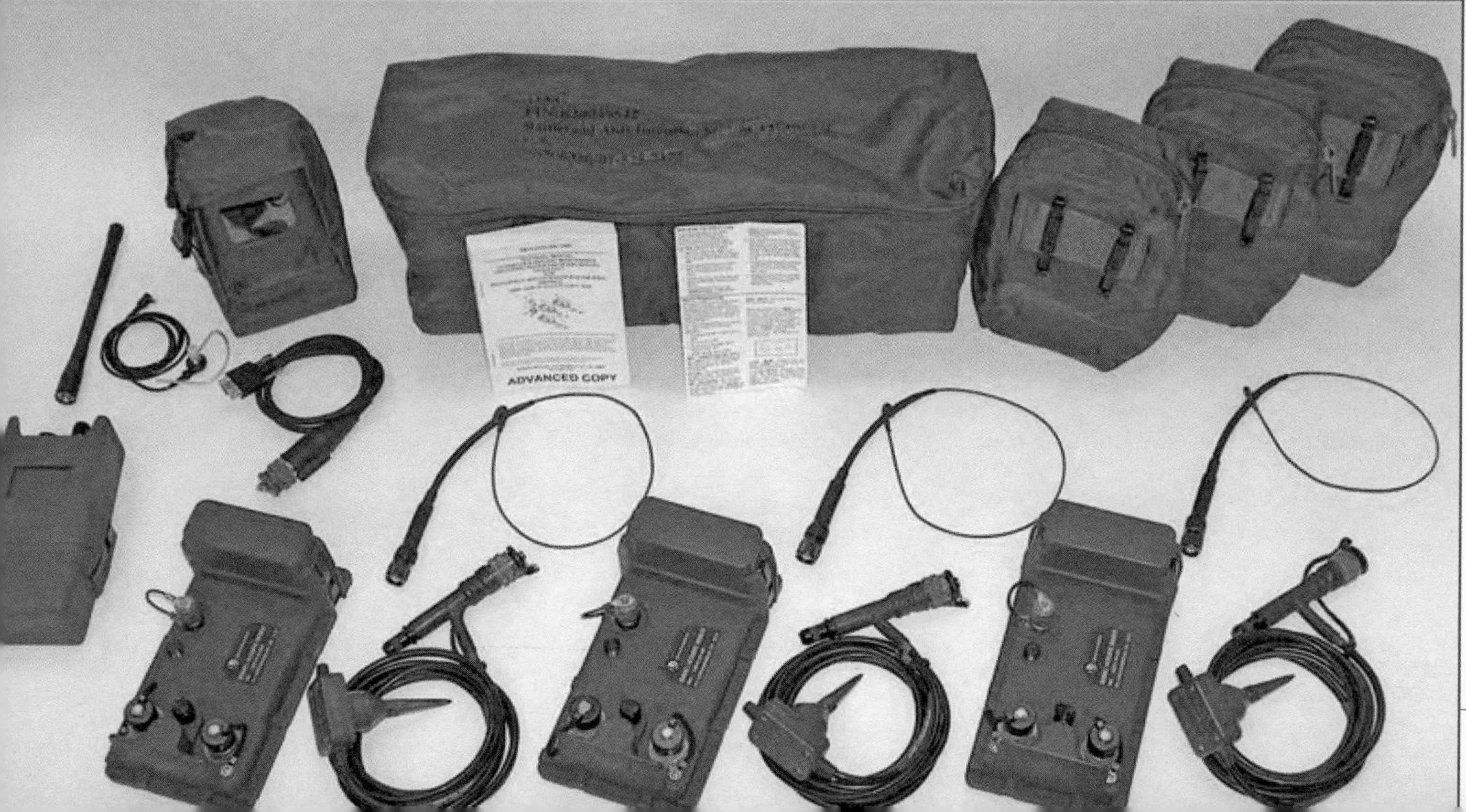

Physical Security Force Protection System of Systems

FOREIGN MILITARY SALES
None

CONTRACTORS
Computer Sciences Corp. (Springfield, VA)
DRS Radian, Inc. (Alexandria, VA)
EG&G Technical Services, Inc. (Albuquerque, NM)
General Dynamics Robotics Systems (Westminster, MD)
L-3 Communications-East (Camden, NJ)

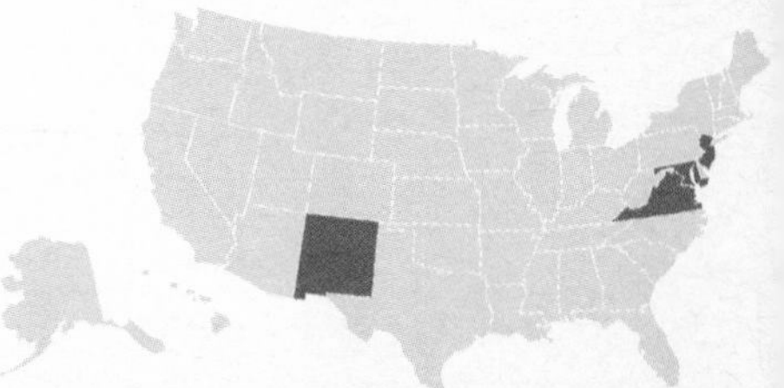

Physical Security Force Protection System of Systems: Automated Installation Entry (AIE)

Provides commanders with the capability to improve security, enhance throughput, and reduce guard requirements/costs for the operation of installation access control point gate operations.

INVESTMENT COMPONENT

- Modernization
- Recapitalization
- Maintenance

Description & Specifications

Automated Installation Entry (AIE) is a software and hardware system designed to read and compare vehicle and personnel identification media. The results of the comparison are used to permit or deny access to installations in accordance with installation commanders' criteria.

AIE will include a database of personnel/vehicles that have been authorized entry onto an Army installation and appropriate entry lane hardware to permit/deny an effort to gain access to the installation. AIE will use current personnel data available from defense personnel databases to validate the authenticity of credentials presented by a person wishing to gain entry and system-generated information from vehicle registration databases pertaining to their defense registered vehicle.

AIE will have the capability to process enrolled visitors as well as permanent personnel and present a denial barrier to restrict the entrance of unauthorized personnel.

AIE will be capable of adapting to immediate changes in threat conditions and apply restrictive entrance criteria consistent with the force protection condition.

Program Status

- **1QFY06:** Access Control Working Group establishes AIE program
- **2QFY06:** AIE Integrated Product Team established to draft service requirements and performance requirements
- **3QFY06:** Draft system specification and performance criteria defined and proof of principle demonstration conducted
- **4QFY06:** Finalize statement of work and performance requirements
- **1QFY07:** Finalize statement of work and performance requirements

Recent and Projected Activities

- **3QFY07:** Award contract
- **4QFY07:** Conduct performance verification testing at pilot installation
- **1QFY08:** Initiate operational evaluation and endurance testing
- **2QFY08:** System installation at first priority base

ACQUISITION PHASE

- Concept & Technology Development
- System Development & Demonstration
- Production & Deployment
- Operations & Support

Physical Security Force Protection System of Systems: Automated Installation Entry (AIE)

FOREIGN MILITARY SALES
None

CONTRACTORS
To be determined

Precision Guided Mortar Munitions (PGMM)

Destroys high-payoff targets, such as earth and timber bunkers, masonry structures, light armored vehicles, and command and control centers with beyond line-of-sight direct fire weapons.

INVESTMENT COMPONENT

- Modernization
- Recapitalization
- Maintenance

Description & Specifications

The 120mm Precision Guided Mortar Munitions (PGMM) is a multi-purpose, laser-guided 120-mm mortar cartridge capable of defeating high-payoff targets with low collateral damage. It will be fired from all U.S. smoothbore 120mm mortars, which is the organic indirect fire asset for all Heavy, Infantry, and Stryker Brigade Combat Teams. After firing, it flies ballistically to its search area, finds the spot being designated by any DoD laser designator (ground, vehicle, or air mounted), and homes in on it. PGMM's warhead will defeat enemy personnel protected by masonry walls, earth and timber bunkers, or within lightly armored vehicles.

Defeating point targets is critical in urban environments and in low- to high-intensity conflicts because it avoids collateral damage and reduces the potential for civilian casualties. PGMM increases the number of stowed kills and reduces the overall logistics burden—a critical goal for early entry forces.

Precision: Two rounds or less to defeat the target (semi-active laser designation for precision strike)
Range: Provides lethality at extended ranges with incremental upgrades (7.2–15 kilometers)
Lethality: Provides high lethality against personnel protected by earth and timber bunkers, lightly armored vehicles, and masonry structures
Shelf Life: 10–20 years
Weapon System: Compatible with all current and future 120mm U.S. mortar systems

Program Status

- **2QFY06:** Conducted Tactics, Techniques and Procedures (TTP) demonstration of the PGMM with the U.S. Army Infantry Center
- **3QFY06:** Demonstrated warhead lethality against all PGMM targets
- **4QFY06:** Demonstrated gun fired closed loop performance (sensor-detected and round-maneuvered target)
- **1QFY07–4QFY07:** PGMM contract development testing (full All-Up-Round [AUR] demonstration)

Recent and Projected Activities

- **4QFY07:** PGMM contract development test completed
- **2QFY07:** Critical design review
- **1QFY08–3QFY08:** Conduct system demonstration phase of system development and demonstration
- **4QFY08:** Milestone C (Increment I)
- **4QFY08–4QFY09:** Low-rate initial production
- **3QFY09–4QFY09:** Production qualification tests and evaluation
- **2QFY10:** Operational test
- **4QFY10:** Materiel release
- **4QFY10:** First unit equipped

ACQUISITION PHASE

Concept & Technology Development | System Development & Demonstration | Production & Deployment | Operations & Support

Precision Guided Mortar Munitions (PGMM)

XM395 Precision Guided Mortar Munition (PGMM) Increment 1

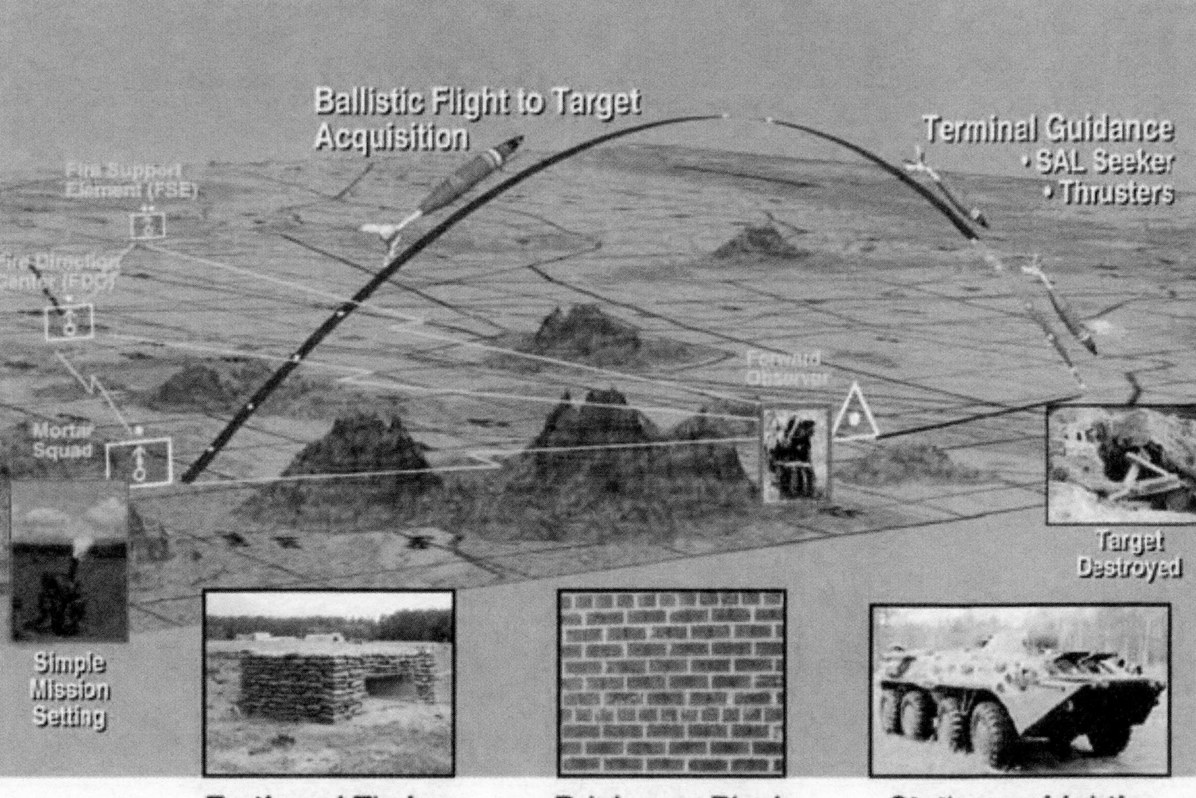

Current 120mm Mortars

Future

Earth and Timber Bunkers	Brick over Block Masonry Structures	Stationary Lightly Armored Vehicles

Increment 1 Targets

FOREIGN MILITARY SALES
None

CONTRACTORS
Prime:
Alliant Techsystems (Plymouth, MN)
Subcontractors:
BAE Systems (Nashua, NH)
Pacific Scientific (Valencia, CA)

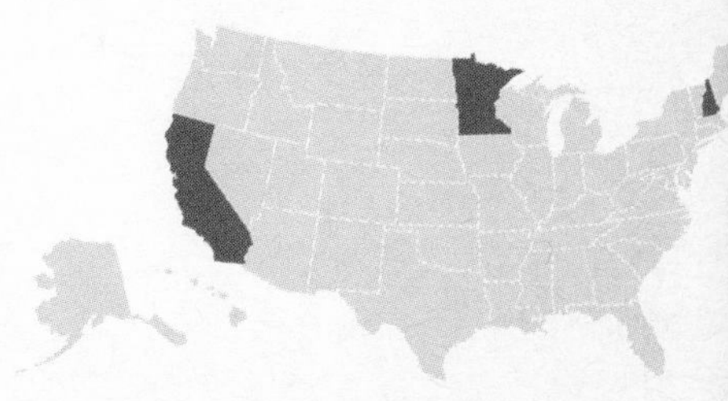

Prophet

Improves battlespace awareness using electronic support sensors that detect, collect, identify, and locate selected emitters, and enhances combat effectiveness using electronic warfare against targeted enemy command and control nodes.

INVESTMENT COMPONENT

- Modernization
- Recapitalization
- Maintenance

Description & Specifications

Providing an electronic support (ES) capability, Prophet creates a near-real-time electronic picture of the brigade/armored cavalry regiment (ACR), Stryker Brigade Combat Team (SBCT), and Brigade Combat Team (BCT) battlespace. Prophet provides intelligence support by reporting the location, tracking, and identity of threat emitters. A secondary mission is electronic attack (EA) against enemy emitters.

The currently fielded Prophet Block I ES system is mounted on a High-Mobility Multipurpose Wheeled Vehicle (HMMWV) with an antenna mast that can be erected quickly. Prophet also has a dismounted man-pack configuration that supports early entry operations. Prophet EA is packaged in a HMMWV trailer towed behind the ES systems, providing both stationary and on-the-move capabilities.

Prophet will cross-cue other battlefield sensors as well as provide additional information that may confirm intelligence from other manned/unmanned battlefield sensors.

Prophet employs an open systems architecture, modular design, and nonproprietary industry standards, supporting evolutionary growth and expansion via circuit card assemblies and software upgrades. This supports the insertion of off-the-shelf technology upgrades to meet theater collection requirements, which has proven effective in Operation Iraqi Freedom and Operation Enduring Freedom through quick-reaction fieldings directly supporting the commander's operational needs. Examples include the Prophet Hammer and Cobra technical insertions.

Prophet is being procured in an evolutionary approach to ensure that enhancements such as reduced footprint and logistics are incorporated.

Program Status

- **3QFY06–1QFY07:** Continued fielding to National Guard and Army Transformation BCTs
- **3QFY06–1QFY07:** Continued production startup of Prophet EA low rate initial production
- **4QFY06:** Engineering Change Proposal awarded for continuation of Prophet ES Spiral 2 (Block III) system development and demonstration
- **1QFY07:** Complete first pre-production Prophet ES Spiral I (Interim Block III) system

Recent and Projected Activities

- **2QFY07–4QFY07:** Complete Prophet Block I fieldings
- **3QFY07:** Operational assessment for Prophet ES Spiral 1 (Interim Block III)
- **4QFY07:** Prophet ES Spiral 1 (Interim Block III) first unit equipped
- **2QFY08:** Prophet EA (Block II) initial operational test and evaluation
- **3QFY08:** Prophet EA (Block II) first unit equipped

ACQUISITION PHASE

- Concept & Technology Development
- System Development & Demonstration
- Production & Deployment
- Operations & Support

Prophet

FOREIGN MILITARY SALES
None

CONTRACTORS

Prophet Block I production:
L-3 Communications (San Diego, CA; Melbourne, FL)

Prophet EA (Block II) production:
General Dynamics C4 Systems, Inc. (Scottsdale, AZ)
Rockwell Collins (Cedar Rapids, IA; White Marsh, MD)

Prophet ES Spiral 1 (Interim Block III) development:
L-3 Communications (San Diego, CA; Melbourne, FL)

Prophet ES Spiral 2 (Block III) Development:
General Dynamics C4 Systems, Inc. (Scottsdale, AZ)

Rapid Equipping Force (REF)

Develops, tests, and evaluates key technologies and systems under operational conditions and delivers them rapidly to operational commanders to increase mission capability while reducing the risk to Soldiers.

INVESTMENT COMPONENT

- Modernization
- Recapitalization
- Maintenance

Description & Specifications

The Rapid Equipping Force (REF) was established to provide operational commanders with rapidly employable solutions to enhance lethality, survivability, and force protection, and to evaluate under operational conditions technologies and systems that support rapid attainment of future force capabilities. REF exploits the full range of possible solutions from the Army, the other services, from within government, and from commerical sources.

REF accomplishes its mission by partnering with the U.S. Army Materiel Command, industry, academia, Army senior leaders, the U.S. Army Training and Doctrine Command, the Army acquisition community, and the Army Test and Evaluation Command to meet immediate warfighter needs.

REF task directly supports the Joint Improvised Explosive Device Defeat Organization and the Army's Asymmetric Warfare Group. In its general support role to the Army, REF forward teams quickly identify and evaluate deployed force needs and desired capabilities. REF develops and rapidly acquires appropriate solutions, while documenting a streamlined methodology for acquisition with the cooperation and oversight of the Army Acquisition Executive.

Program Status

- **4QFY06:** To date REF has introduced 209 technologies, placing more than 47,000 items of equipment in Soldiers' hands

Recent and Projected Activities

- **FY07:** REF will continue to provide immediate support to warfighters in while expanding operations worldwide
- **FY08:** REF will enter the Army base program and budget with reduced reliance on supplemental funding

ACQUISITION PHASE

Concept & Technology Development | System Development & Demonstration | Production & Deployment | Operations & Support

Alternative Power Source

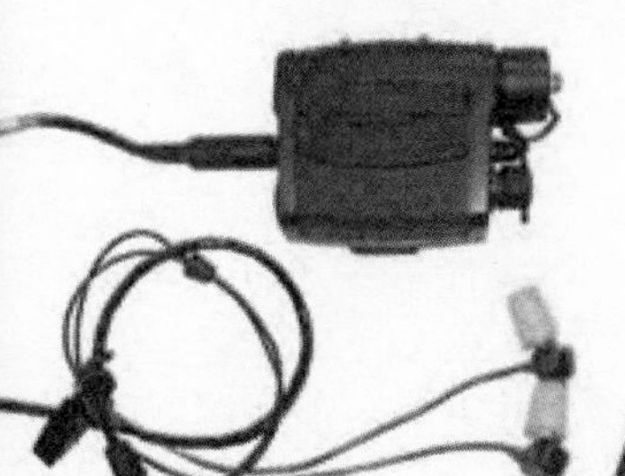

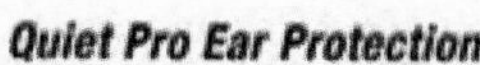

Quiet Pro Ear Protection

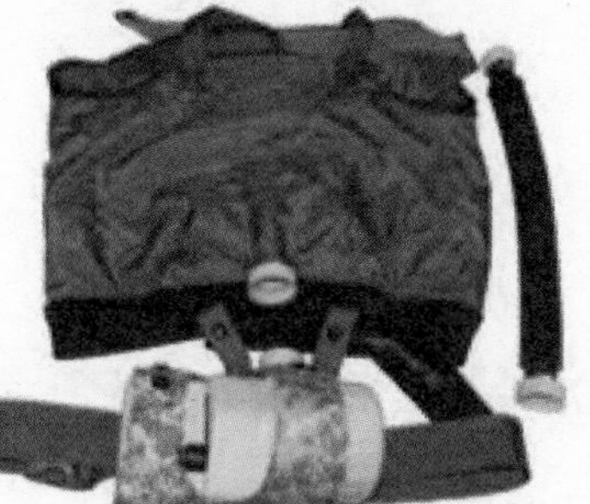

Body Ventilation System

FN303 Non-Lethal Projectile Launcher

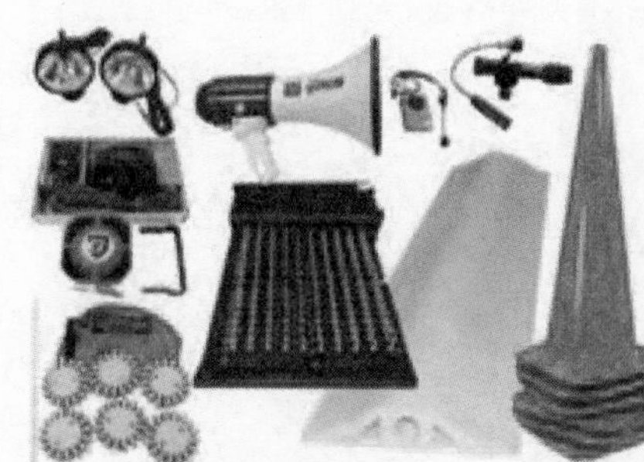

Escalation of Force

Rapid Equipping Force (REF)

FOREIGN MILITARY SALES
None

CONTRACTORS
General Atomics (San Diego, CA)
Global Secure Corp. (Washington, DC)
Buffalo Turbine LLC (Springville, NY)
Gyrocam Systems (Sarasota, FL)
Tactical Support Equipment (Fayetteville, NC)

Secure Mobile Anti-Jam Reliable Tactical-Terminal (SMART-T)

Achieves end-to-end connectivity that meets joint requirements for command, control, communications, computers, and intelligence.

INVESTMENT COMPONENT

Modernization

Recapitalization

Maintenance

Description & Specifications

The Secure Mobile Anti-Jam Reliable Tactical Terminal (SMART-T) is a mobile military satellite communication terminal mounted on a standard High Mobility Multipurpose Wheeled Vehicle (HMMWV). The SMART-T extends the range for current and future tactical communications networks through DoD Milstar communication satellites.

SMART-T's maximum rate for data and voice communications is 1.544 million bits per second (Mbps). It provides the Army its only protected (anti-jam) wideband satellite communication capability. Development is under way to upgrade terminals to communicate with DoD Advanced Extremely High Frequency (AEHF) satellites at a maximum data rate of 8.192 Mbps.

Program Status

- **2QFY05–4QFY06:** Fielded 68 SMART-T medium data rate terminals to Army and Marine Corps units
- **2QFY05–4QFY06:** Deployed up to 61 SMART-Ts in Southwest Asia; received positive feedback
- **3QFY04–1QFY05:** Continued AEHF development upgrade effort to provide the warfighter with increased data rates
- **FY06:** Procured 326 SMART-Ts and fielded 89 terminals (all services)

Recent and Projected Activities

- **2QFY07:** Award production contract for AEHF upgrade kits
- **1QFY07–4QFY08:** Complete fielding of SMART-Ts, including 49 to Army National Guard units
- **1QFY09:** Begin installation of AEHF upgrade kits to all SMART-Ts

ACQUISITION PHASE

Concept & Technology Development | System Development & Demonstration | Production & Deployment | Operations & Support

Secure Mobile Anti-Jam Reliable Tactical-Terminal (SMART-T)

FOREIGN MILITARY SALES

A memorandum of understanding has been signed with Canada, the United Kingdom, and the Netherlands for the development, production, and operational and support phase of the AEHF satellite program. A design review for a version of the SMART-T for these international partners was held 1QFY07.

CONTRACTORS

AEHF development:
Raytheon (Marlborough, MA)
Production:
Raytheon (Largo, FL)
Engineering support:
Lincoln Labs (Lexington, MA)
Hardware:
Sechan Electronics (Lititz, PA)
Administative/technical support:
JANUS Research (Eatontown, NJ)

Sentinel

Provides critical air surveillance by automatically detecting, tracking, classifying, identifying, and reporting targets to air defense weapons systems and battlefield commanders.

INVESTMENT COMPONENT

- Modernization
- Recapitalization
- Maintenance

Description & Specifications

Sentinel is used with the Army's Forward Area Air Defense Command and Control (FAAD C2) system and provides key target data to Stinger-based weapon systems and battlefield commanders via FAAD C2 or directly, using an Enhanced Position Location Reporting System (EPLRS) or the Single Channel Ground and Airborne Radio System.

Sentinel consists of the M1097A1 High Mobility Multipurpose Wheeled Vehicle (HMMWV), the antenna transceiver group mounted on a high mobility trailer, the identification friend-or-foe system (IFF), and the FAAD C2 interface. The sensor is an advanced three-dimensional battlefield X-band air defense phased-array radar with a range of 40 kilometers.

Sentinel can operate day and night, in adverse weather conditions, and in battlefield environments of dust, smoke, aerosols, and enemy countermeasures. It provides 360-degree azimuth coverage for acquisition and tracking of targets (cruise missiles, unmanned aerial vehicles, rotary, and fixed wing aircraft) moving at supersonic to hovering speeds and at positions from the nap of the earth to the maximum engagement altitude of short-range air defense weapons. Sentinel picks up targets before they can engage, thus improving air defense weapon reaction time and allowing engagement at optimum ranges. Sentinel's integrated IFF system reduces the potential for engagement of friendly aircraft.

Sentinel modernization efforts include enhanced target range and classification (ETRAC) upgrades to engage non-line of sight targets; increased detection and acquisition range of cruise missiles, unmanned aerial vehicles and fixed/rotary wing targets; enhanced situational awareness; and classification of cruise missiles. The system provides integrated air tracks with classification and recognition of platforms that give an integrated air and cruise missile defense solution for the Air and Missile Defense System of Systems Increment 1 architecture and subsequent increments.

Sentinel has been critical in providing air surveillance of the National Capital Region and other areas as a part of ongoing homeland defense efforts.

Program Status

- **3QFY06:** Initial Sentinel ETRAC system retrofit of fielded system
- **4QFY06:** Sentinel ETRAC initial first unit equipped

Recent and Projected Activities

- **1QFY08:** Sentinel joint ID prototype
- **3QFY08:** Mode 5/S IFF demonstration

ACQUISITION PHASE

Concept & Technology Development | System Development & Demonstration | Production & Deployment | Operations & Support

Sentinel

FOREIGN MILITARY SALES
None

CONTRACTORS
Thales Raytheon Systems (Fullerton, CA; Forest, MS; Largo, FL)

Small Arms

Enables warfighters and small units to engage targets with lethal fire to defeat or deter adversaries.

INVESTMENT COMPONENT

Modernization

Recapitalization

Maintenance

Description & Specifications

The M4 Carbine is a compact version of the M16A2 rifle, with a collapsible stock, a flat-top upper receiver accessory rail, and a detachable handle/rear sight aperture sight assembly. It achieves more than 85 percent commonality with the M16A2 rifle and replaces all .45 caliber M3 submachine guns, selected M9 pistols, and M16 series rifles.

The M249 Squad Automatic Weapon (SAW) is a lightweight, gas-operated, one-man-portable automatic weapon that delivers substantial, effective fire at ranges out to 1,000 meters. Improved bipods, improved collapsible buttstocks, lightweight ground mounts, and improved combat optics have increased combat effectiveness of this weapon.

The **M240 Series of Machine Gun** is the ground version of the original M240 machine gun. The M240H is used as a defensive armament for the UH-60 Black Hawk and CH-47 Chinook. A lighter weight M240E6 will replace the M240B in Special Force/Ranger, Light Infantry, and Airborne units.

The MK19, Mod 3 Grenade Machine Gun is self-powered and air-cooled. It engages point targets up to 1,500 meters and provides suppressive fires up to 2200 meters. It can be mounted on various tracked and wheeled vehicles, and on the M3 tripod for static defensive operations.

The XM320 Grenade Launcher Module attaches to the M4 Carbine and M16A1/M16A2 Rifles and fires all existing and improved 40mm low-velocity ammunition. It can also be configured as an independent weapon.

The XM26 Modular Accessory Shotgun System attaches to the M4 carbine and M16A2/M16A4 rifles and fires all standard lethal, non-lethal, and door-breaching 12 gauge ammunition. It can also be configured as an independent weapon.

Program Status

Various

Recent and Projected Activities

M4 Carbine:
- Continue production and fielding

M249 SAW:
- Continue production and fielding

M240 Machine Gun:
- Continue production and fielding

Mk19 Grenade Machine Gun:
- Continue production and fielding

XM320 Grenade Launcher Module:
- **2QFY07:** Milestone C
- **4QFY07:** Operational test
- **1QFY08:** Full rate production decision
- **2QFY08:** First unit equipped

XM26 Modular Accessory Shotgun System:
- **2QFY07:** Milestone C
- **1QFY08:** Full rate production
- **2QFY08:** First unit equipped

ACQUISITION PHASE

Concept & Technology Development | System Development & Demonstration | Production & Deployment | Operations & Support

Small Arms

FOREIGN MILITARY SALES
Numerous foreign countries purchase U.S. small arms

CONTRACTORS
M4 Carbine:
Colt's Manufacturing (Hartford, CT)
M249 SAW:
Fabrique National Manufacturing, LLC (Columbia, SC)
M240B Machine Gun:
Fabrique National Manufacturing, LLC (Columbia, SC)
Mk19 Grenade Machine Gun:
General Dynamics ATP Division (Saco, ME)
XM320 Grenade Launcher Module:
Heckler and Koch Defense Inc. (Sterling, VA)
XM26 Modular Accessory Shotgun System: Vertu Corp. (Manassas, VA)

Small Caliber Ammunition

Provides the highest quality small caliber ammunition to warfighters for training and combat.

INVESTMENT COMPONENT

- Modernization
- Recapitalization
- Maintenance

Description & Specifications

The Small Caliber Ammunition program consists of the following cartridges: 5.56mm, 7.62mm, 9mm, 10-gauge and 12-gauge shotgun, .22 caliber, .30 caliber, and .50 caliber. Small Caliber Ammunition supports the following Soldier weapons: M9 pistol, M16A1/A2/A4 rifle, M4 carbine, M249 squad automatic weapon, M240 machine gun, .50-caliber M2 machine gun, sniper rifles, and a variety of shotguns. The .30 caliber blank cartridge supports Veterans Service Organizations performing veterans' funeral honors.

Program Status

- In production

Recent and Projected Activities

- **2QFY07:** Lake City Army Ammunition Plant capacity expansion achieves 1.6 billion rounds
- **2QFY07:** 5.56mm lead replacement program low rate initial production decision

ACQUISITION PHASE

Concept & Technology Development | System Development & Demonstration | Production & Deployment | Operations & Support

Small Caliber Ammunition

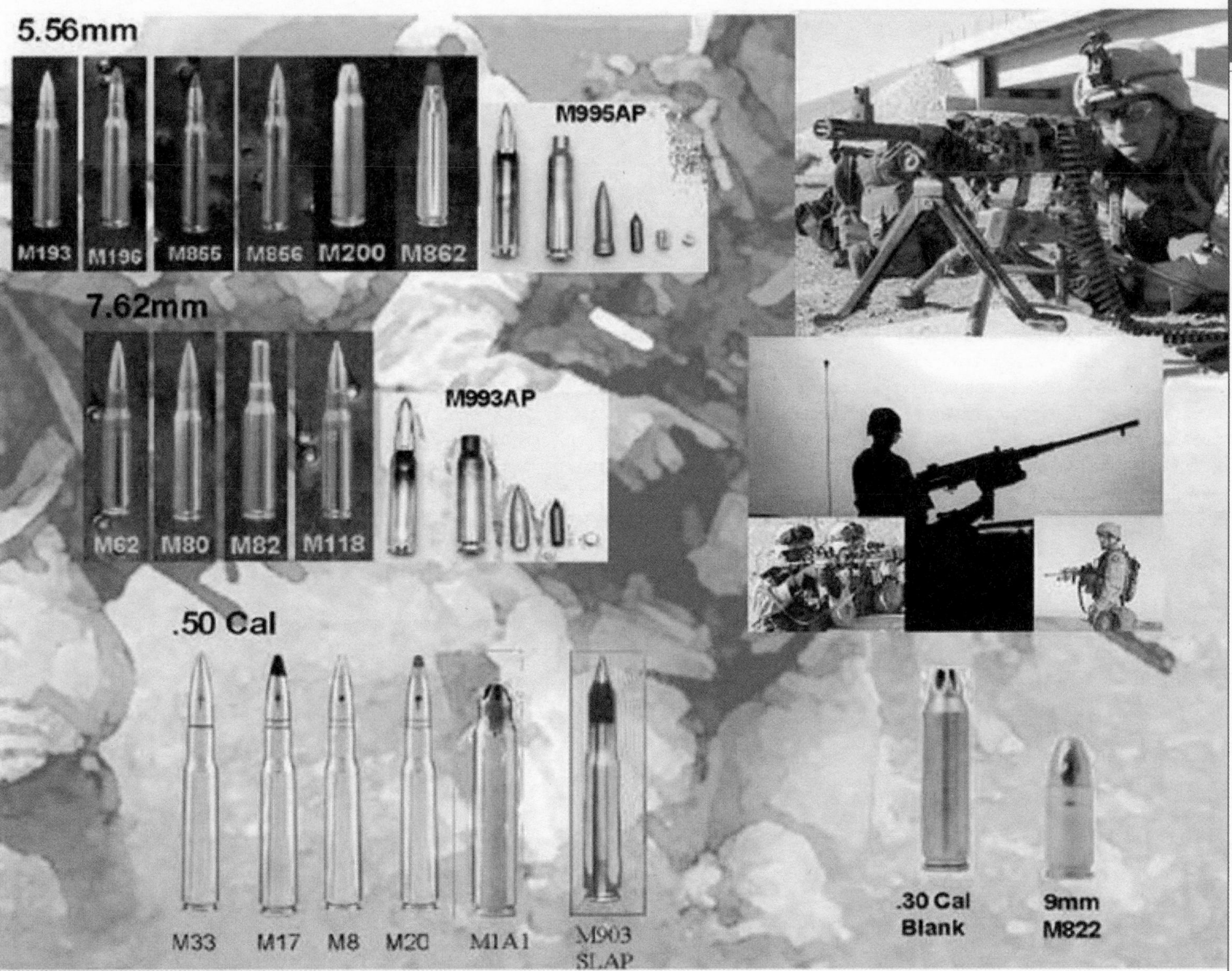

FOREIGN MILITARY SALES

5.56mm: Japan, Philippines
7.62mm: Antigua and Barbuda, Colombia, Japan, Jordan, Philippines

CONTRACTORS

Alliant Techsystems (Independence, MO)
General Dynamics Ordnance and Tactical Systems (St. Petersburg, FL)
Nordic Ammunition (Karlsborg, Sweden)

Sniper Systems

Enables sniper teams to engage targets at extended ranges with lethal force to defeat or deter adversaries.

INVESTMENT COMPONENT

- Modernization
- Recapitalization
- Maintenance

Description & Specifications

Sniper systems consist of the following two weapons and their accessories:

M107 Semi-Automatic Long Range Sniper Rifle (LRSR) is a commercial off-the-shelf anti-materiel and counter sniper semi-automatic .50 caliber rifle. The rifle is a reliable, direct-line-of-sight weapon system, capable of delivering precise rapid fire on targets out to 2000 meters.

The XM110 Semi-Automatic Sniper System (SASS) is a commercial off-the-shelf, anti-personnel, 7.62mm semi-automatic sniper rifle that is also effective against light materiel targets. Capable of rapid fire/rapid reload, this suppressed sniper rifle exceeds the rate-of-fire and lethality of the M24 Sniper Weapon System. SASS anti-personnel ranges are equal to or greater than the M24. SASS includes an enhanced sniper spotting scope.

Program Status

- **M107:** In production; maintenance work order/upgrade under way
- **XM110:** Fielding in support of urgent materiel releases

Recent and Projected Activities

- **M107:** Continue planned procurements
- **1QFY07:** Complete Milestone C for XM110
- **2QFY07:** Begin limited rate initial production of XM110

ACQUISITION PHASE

- Concept & Technology Development
- System Development & Demonstration
- Production & Deployment
- Operations & Support

Sniper Systems

FOREIGN MILITARY SALES
None

CONTRACTORS
M107:
Barrett Firearms Manufacturing (Murfreesboro, TN)
XM110:
Knight's Armaments Co. (Titusville, FL)

Stryker

Enables deployment of Stryker Brigade Combat Teams (SBCTs) anywhere in the world using readily deployable, combat-ready support vehicles capable of rapid movement.

INVESTMENT COMPONENT

Modernization

Recapitalization

Maintenance

Description & Specifications

Stryker is a family of eight-wheeled armored vehicles that combine high battlefield mobility, firepower, survivability, and versatility with reduced logistics requirements. It includes two types of vehicles: the Infantry Carrier Vehicle (ICV) and the Mobile Gun System (MGS). The ICV, a troop transport vehicle, can carry nine infantry Soldiers, their equipment, and a crew of two: driver and vehicle commander. The MGS, designed to support infantry, has a 105mm turreted gun and autoloader system to breach bunkers and concrete walls.

Eight other configurations based on the ICV support combat capabilities: Reconnaissance Vehicle (RV), Mortar Carrier (MC), Commander's Vehicle (CV), Fire Support Vehicle (FSV) Engineer Squad Vehicle (ESV), Medical Evacuation Vehicle (MEV), Anti-Tank Guided Missile (ATGM) Vehicle, Nuclear, Biological, and Chemical Reconnaissance Vehicle (NBCRV)

The ICV (excluding the MEV, ATGM, FSV, and RV) is armed with a remote weapons station supporting an M2 .50-caliber machine gun or MK19 automatic grenade launcher, the M6 grenade launcher, and a thermal weapons sight. Stryker supports a communications suite integrating the Single Channel Ground and Airborne Radio System (SINGCARS); Enhanced Position Location Reporting System (EPLRS); Force XXI Battle Command Brigade-and-Below (FBCB2); GPS; and high-frequency and near-term digital radio systems. In urban terrain, Stryker gives 360-degree protection against 14.5mm armor piercing threats. It is deployable by C-130 aircraft and combat-capable on arrival.

The Stryker program leverages non-developmental items with common subsystems and components to quickly field these systems. Strykers integrate government furnished materiel subsystems as necessary. Stryker stresses performance and commonality to reduce the logistics footprint and minimize costs. Since October 2003, Strykers in Iraq have logged over 7 million miles and kept operational readiness above 90 percent.

SBCTs require 328 Stryker variants, because of an added Stryker-based retrans and gateway capability. The current program requires over 3,100 Strykers to field seven SBCTs and meet additional requirements. Funding has increased Stryker Ready to Fight fleet requirements to 150 Strykers, and Repair Cycle Floats to 123 Strykers.

Program Status

- **2QFY05–4QFY06:** Completed fielding to SBCT 4; began fielding to SBCT 5 and SBCT 6
- **1QFY07:** NBCRV independent operational test and evaluation

Recent and Projected Activities

- **2QFY07:** MGS initial operational test and evaluation
- **4QFY07:** MGS and NBCRV MS III

ACQUISITION PHASE

Concept & Technology Development | System Development & Demonstration | Production & Deployment | Operations & Support

Stryker

FOREIGN MILITARY SALES
ICVs: Israel

CONTRACTORS
General Dynamics (Anniston, AL; Sterling Heights, MI; Lima, OH)

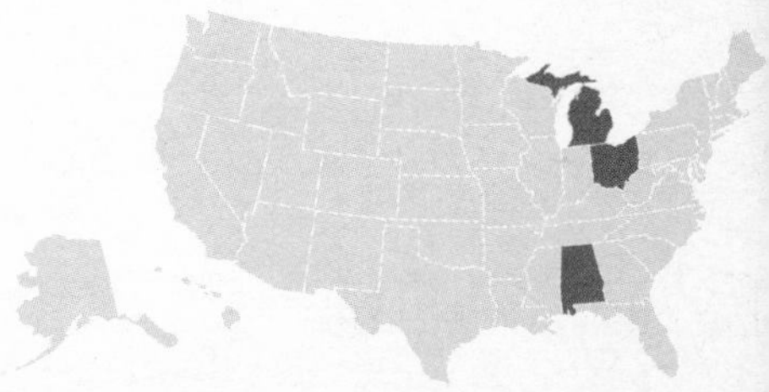

Surface Launched Advanced Medium Range Air-to-Air Missile (SLAMRAAM)

Provides networked air and missile defense capability for the maneuver force, critical geopolitical assets, and homeland defense.

INVESTMENT COMPONENT

- Modernization
- Recapitalization
- Maintenance

Description & Specifications

The Surface Launched Advanced Medium Range Air-to-Air Missile (SLAMRAAM) will defend designated critical assets and maneuver forces against aerial threats. It is a key component of the Integrated Air and Missile Defense (IAMD) Composite Battalion and will replace the Avenger in the Army's Air and Missile Defense forces. SLAMRAAM is a lightweight, day-or-night, and adverse weather, non-line of sight system for countering cruise missiles and unmanned air vehicle threats with engagement capabilities in excess of 18 kilometers. The system comprises an Integrated Fire Control Station (IFCS) for command and control, integrated sensors, and missile launcher fire unit platforms. While SLAMRAAM uses its own Sentinel Enhanced Target Range Acquisition Classification (ETRAC) radar to provide surveillance and fire control data, the system will receive data from other joint and Army external sensors when available. The SLAMRAAM launcher is a mobile platform with common joint launch rails, launcher electronics, on-board communication components, and four to six AIM-120-C7 Advanced Medium Range Air-to-Air Missiles (AMRAAMs).

Program Status

- **3QFY06:** System critical design review

Recent and Projected Activities

- **4QFY07:** Delivery of Air and Missile Defense common software build
- **FY08:** System development testing

ACQUISITION PHASE

- Concept & Technology Development
- System Development & Demonstration
- Production & Deployment
- Operations & Support

Surface Launched Advanced Medium Range Air-to-Air Missile (SLAMRAAM)

FOREIGN MILITARY SALES
None

CONTRACTORS
Raytheon (Tewksbury, MA)
Boeing (Huntsville, AL)

Tactical Electric Power (TEP)

Provides modernized tactical electric power sources for all services.

INVESTMENT COMPONENT

Modernization

Recapitalization

Maintenance

Description & Specifications

The Tactical Electric Power (TEP) program consists of small (0.5kW) to large (840kW) generating systems and power units and power plants providing "single fuel" (diesel/JP-8) electrical generator systems that:

- Increase reliability (now 500–600 hours mean time between failure)
- Reduce weight/cube
- Reduce infrared signature and noise (to 70 dBA [decibels] at 7 meters)
- Are survivable in chemical, biological, and nuclear environments
- Provide quality electric power for command posts; command, control, communications, computers, intelligence, surveillance, and reconnaissance (C4ISR) systems; weapon systems; and other battlefield support equipment.

Program Status

- **FY06:** Production and fielding ongoing for 2kW (kilowatt) Military Tactical Generator (MTG), 3kW, 5kW, 10kW, 15kW, 30kW, 60kW Tactical Quiet Generator (TQG), and 840kW Deployable Power Generation and Distribution System (DPGDS) generator sets
- **FY06:** Production ongoing for 100kW and 200kW TQG
- **FY06:** Continued assembly and fielding of power units and power plants (trailer-mounted generator sets)
- **FY06:** Advanced Medium Mobile Power Sources (AMMPS) (next-generation TEP sources) Phase 1 completed (multiple contractors build and test prototypes; government evaluation)
- **1QFY07:** First production deliveries and fielding of 100kW and 200kW generator sets
- **1QFY07:** AMMPS Phase 2 design contract award (build pre-production units and conduct pre-production qualification test (PPQT))

Recent and Projected Activities

- **FY07:** Continue production and fielding of generator sets and power units and power plants
- **3QFY07:** Award Power Distribution Illumination System Electric (PDISE) competitive contract

ACQUISITION PHASE

Concept & Technology Development | System Development & Demonstration | Production & Deployment | Operations & Support

Tactical Electrical Power (TEP)

2kW Military Tactical Generator

3kW Tactical Quiet Generator

30kW Tactical Quiet Generator

5kW Tactical Quiet Generator

10kW Tactical Quiet Generators

15kW Tactical Quiet Generator

60kW Tactical Quiet Generator

100kW Tactical Quiet Generator

840kW Deployable Power Generation and Distribution System

FOREIGN MILITARY SALES
Tactical quiet generators have been purchased by Egypt, Israel, Korea, Kuwait, Saudi Arabia, Turkey, United Arab Emirates, and 11 other countries

CONTRACTORS
3kW, 5kW, 10kW, 15kW, 100kW and 200kW TQG:
DRS Fermont (Bridgeport, CT)
30kW, 60kW TQG:
MCII (Tulsa, OK)
840kW DPGDS:
DRS-ESSI (White Plains, MO)
2kW MTG:
Dewey Electronics (Oakland, NJ)
Trailers for power units and power plants: Phoenix Coaters LLC (Berlin, WI)

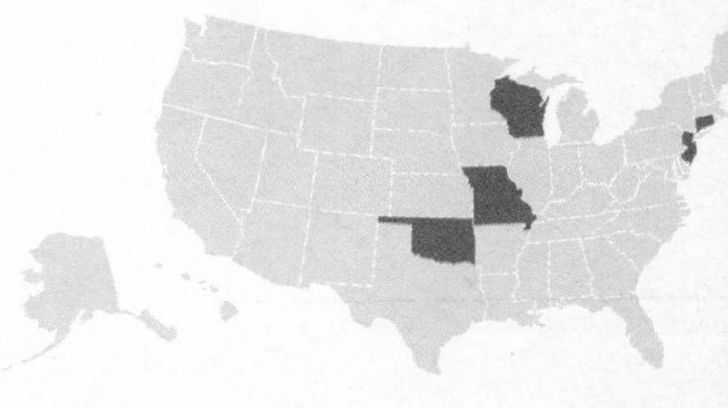

Tactical Fire Fighting Truck (TFFT)

Supports the tactical fire fighting team with a mobile, versatile vehicle capable of performing multiple missions.

INVESTMENT COMPONENT

- Modernization
- Recapitalization
- Maintenance

Description & Specifications

The multi-purpose Tactical Fire Fighting Truck (TFFT), issued to tactical firefighting detachments and ammunition companies, is used primarily for aircraft and vehicle rescue, limited structural fires, hazardous material containment, and wild fire control. It is the primary vehicle of the Tactical Fire Fighting Team, which consists of the TFFT, two 1750-gallon water distribution modules, one Heavy Expanded Tactical Truck (HEMTT) with Load Handling System, and one Palletized Load System trailer. The HEMTT chassis enables greater mobility and all-terrain maneuverability. Parts commonality with HEMTT reduces the logistics footprint.

The TFFT accommodates a six-person crew (five in full SCBA [self-contained breathing apparatus]). It is equipped with an around-the-pump twin-agent foam system, a 1000-gallon water tank, 2000 feet of hose, a 1000-gallon per minute pump, and roof and bumper turrets with in-cab remote controls. Storage compartments carry everything needed on scene, including a hydraulic generator, rescue tools, saws, air-lifting bag kit, extrication equipment, and EMS equipment.

Truck gradability: 60 percent
Maximum speed: 62 mph
Flatrack dimensions: 8 x 20 foot (ISO container standard)
Engine type: Detroit diesel model 8V92TA/445-450 hp, 12.1 liter DDEC
Transmission: Allison HD 4560, Automatic
Cruising range: 400 miles
Fording clearance: 48 inches
Number of driven wheels: 8
Air transportability: C-17

Program Status

Current: Production and deployment

Recent and Projected Activities

Fielding

ACQUISITION PHASE

Concept & Technology Development | System Development & Demonstration | Production & Deployment | Operations & Support

Tactical Fire Fighting Truck (TFFT)

FOREIGN MILITARY SALES
None

CONTRACTORS
Oshkosh Truck Corp. (Pierce)
(Appleton, WI)
United Plastics Fabricators (Neenah, WI)
W. S. Darley Corp. (Melrose Park, IL)
Deutz U.S.A (Atlanta, GA)
Akron Brass (Wooster, OH)

Tactical Operations Center (TOC)

Provides commanders at standardized and mobile command posts with a tactical, fully integrated, and digitized physical infrastructure to execute battle command and achieve information dominance.

INVESTMENT COMPONENT

- **Modernization**
- Recapitalization
- Maintenance

Description & Specifications

The Tactical Operations Center (TOC) program provides the commander and his staff with a digitized platform and command information center to plan, direct, and track operations.

TOC is based on the Standard Integrated Command Post System (SICPS) family of systems, which includes the SICPS Command Post Platform (CPP), Trailer Mounted Support Systems (including tents, environmental control, power management, and lighting), Command Center System (CCS), Command Post Communications System (CPCS), and Command Post Local Area Network (CP LAN). SICPS integrates approved and fielded command and control and other command, control, communications, computers, intelligence, and reconnaissance (C4ISR) systems technology into platforms supporting the needs of the current Mechanized, Light, and Stryker Brigade Combat Team (SBCT) forces.

SICPS allows commanders and staffs to digitally plan, prepare, and execute network centric operations through visualization of the common operational picture (COP) and shared situational awareness. SICPS provides a means to host the Army Battle Command System Information Services (AIS) server associated with the ABCS 6.4 architecture, as well as Maneuver Control System Sequential Query Language (MCS SQL) servers, Command Post of the Future (CPOF) servers, and servers associated with GCCS-A and in the future JC2-Army.

Program Status

- **4QFY05:** SICPS Milestone C low rate initial production approval
- **3QFY06:** First unit equipped fielding to 1 Cavalry Division
- **3QFY06:** Conducted successful intitial operational test and evaluation at National Training Center
- **4QFY06:** Conducted successful logistics/maintenance demonstration at Tobyhanna Army Depot
- **4QFY06:** Completed SICPS fielding to SBCT 7
- **4QFY06:** Completed SICPS fielding to two BCTs and headquarters of the 82nd Airborne
- **1QFY07:** Demonstrate mobile-CP LAN prototype to address HMMWV-armor weight issues
- **1QFY07:** Complete SICPS fielding to 82nd Airborne
- **1QFY07:** SICPS full rate production decision

Recent and Projected Activities

- **2QFY07:** Integrate new Battle Command Server stack
- **2QFY07:** Complete SICPS fielding to SBCT 5
- **2QFY07:** Complete SICPS fieldings for 173 AB BCT and 12 CAB (Europe)
- **3QFY07:** Complete SICPS fielding for 3 ID
- **3–4QFY07:** Field 101 AA and reset 4 ID

ACQUISITION PHASE

- Concept & Technology Development
- System Development & Demonstration
- **Production & Deployment**
- Operations & Support

Tactical Operations Center (TOC)

FOREIGN MILITARY SALES
None

CONTRACTORS
Hardware design and integration (SICPS CPP and CCS):
Northrop Grumman (Huntsville, AL)

Tactical Unmanned Aerial Vehicle (TUAV)

Provides the tactical maneuver commander near real-time reconnaissance, surveillance, target acquisition, and force protection during day/night and adverse weather conditions.

INVESTMENT COMPONENT

- Modernization
- Recapitalization
- Maintenance

Description & Specifications

The RQ-7B Shadow Tactical Unmanned Aerial Vehicle (TUAV) has a wingspan of 14 feet and a payload capacity of approximately 60 pounds; gross takeoff weight is more than 380 pounds and endurance is more than six hours on-station at a distance of 50 kilometers. The system is compatible with the All Source Analysis System, Advanced Field Artillery Tactical Data System, Joint Surveillance Target Attack Radar System Common Ground Station, Joint Technical Architecture-Army, and the Defense Information Infrastructure Common Operating Environment. The One System Ground Control Station (OSGCS) is also the only joint-certified GCS in the DOD. The RQ-7B Shadow can be transported by three C-130 transports.

The RQ-7B Shadow configuration, fielded in platoon sets, consists of:

- Four air vehicles with electro-optic/infrared imaging payloads including infrared illuminators
- Two GCS shelters mounted on High Mobility Multipurpose Wheeled Vehicles (HMMWV) and their associated ground data terminals; one portable GCS and one portable ground data terminal
- One air vehicle transport HMMWV towing a trailer-mounted hydraulic launcher
- Two HMMWV with trailers for operations/maintenance personnel and equipment transport
- One HMMWV with Maintenance Section Multifunctional (MSM) shelter & trailer
- One HMMWV with Mobile Maintenance Facility (MMF) shelter
- Two Automatic Take-off and Landing Systems (TALS)
- Four One System Remote Video Terminals (OSRVT) and antennas

The Shadow is manned by a platoon of 22 soldiers and, typically, two contractors. The soldier platoon consists of a platoon leader, platoon sergeant, UAV warrant officer, 12 Air Vehicle Operators (AVO)/Mission Payload Operators (MPO), four electronic warfare repair personnel and three engine mechanics supporting launch and recovery. The MSM is manned by Soldiers, transporting spares and providing maintenance support. The MMF is manned by contractor personnel located with the Shadow platoon to provide logistics support to include "off system support" and "maintenance by repair."

The Shadow also has an early entry configuration of 15 Soldiers, one GCS, the air vehicle transport HMMWV, and the launcher trailer, which can be transported in one C-130. All components can be slung under a CH-47 or CH-53 helicopter for transport.

Program Status

- Currently fielding to Active and Reserve Component Brigade Combat Teams (BCT). A major acquisition success story, Shadow went from Milestone B to full-rate production decision in just 33 months. Over 48 total systems have been fielded and systems continue to support ground forces in Operation Iraqi Freedom (OIF) and Operation Enduring Freedom (OEF)
- As of FY06 Shadow has flown more than 23,200 sorties and more than 105,600 hours in support of OIF ground forces
- **FY07:** Nine systems on the fielding schedule, with priority to OIF-bound units

Recent and Projected Activities

- **FY07–FY09** Continue fielding shadow platoons in support of Army Modularity, Integrate Tactical Common Data Link (TCDL) and Laser Designation systems; potential for Heavy Fuel Engine (HFE) retrofit

ACQUISITION PHASE

- Concept & Technology Development
- System Development & Demonstration
- Production & Deployment
- Operations & Support

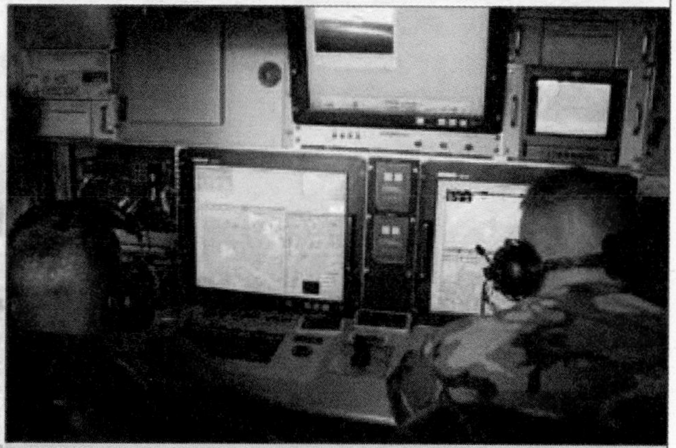

Tactical Unmanned Aerial Vehicle (TUAV)

FOREIGN MILITARY SALES
Pending

CONTRACTORS
Air Vehicle/Ground Data Terminal:
AAI Corp. (Hunt Valley, MD)
GCS, Portable GCS:
CMI (Huntsville, AL)
Auto-land system:
Sierra Nevada Corp (Sparks, NV)
Ground Data Terminal Pedestal:
Tecom (Chatsworth, CA)
MMF/MSM shelter:
General Dynamics (Marion, VA)
Training and tech manuals:
DPA (Arlington, VA)

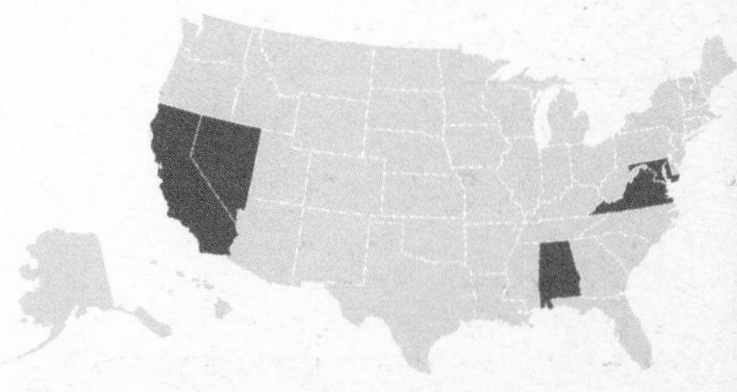

Tank Ammunition

Provides direct-fire tank ammunition for use in current and future ground combat weapons platforms.

INVESTMENT COMPONENT

Modernization

Recapitalization

Maintenance

Description & Specifications

The current 120mm family of tank ammunition consists of kinetic energy, multipurpose and canister ammunition.

Kinetic energy ammunition lethality is optimized by firing a maximum-weight projectile at the greatest velocity possible. The M829A3 is the only kinetic energy cartridge currently in production. Multipurpose ammunition uses a high-explosive, shaped-charge warhead to provide blast, armor penetration, and fragmentation effects. There are three high-explosive cartridges in the current inventory: M830A1, M830 and M908 Obstacle Reduction.

The shotgun shell-like M1028 canister cartridge provides the Abrams tank with effective, rapid, lethal fire against massed assaulting infantry.

To support the Stryker Force, 105mm Mobile Gun System ammunition comprises new high-explosive and canister cartridges. High-explosive ammunition (M393A3) destroys hardened enemy bunkers and creates openings through which infantry can pass. The M1040 canister cartridge provides rapid, lethal fire against massed assaulting infantry at close range.

For Future Combat Systems, smart, precision munitions will enable precision strikes against high-value targets at extended ranges. The mid range munition (MRM) will expand the future combat systems engagement zone beyond 8 kilometers.

Program Status

- **FY06:** M829A3, M830, M830A1 and M908 fielded; M1028 urgent fielded to support Operation Iraqi Freedom; low rate initial production for M1002, M1028, M1040

Recent and Projected Activities

- **FY07:** Material release for M1002; Milestone B and entry into system development and design for MRM

ACQUISITION PHASE

Concept & Technology Development | System Development & Demonstration | Production & Deployment | Operations & Support

MRM Concept

105mm for MGS

120mm Tactical and Training

FOREIGN MILITARY SALES
M829: Kuwait, Saudi Arabia
M830: Kuwait, Egypt, Israel
KE-W/A1 and KE-W/A2: Egypt

CONTRACTORS
M830A1, M1002 and M1040:
Alliant Techsystems (Plymouth, MN)
M1028:
General Dynamics-Ordnance and Tactical Systems (St. Petersburg, FL)
M829A3:
Alliant Techsystems (Plymouth, MN; Rocket City, WV)
AOT (Johnson City, TN)
MRM:
Alliant Techsystems (Clearwater, FL)
Raytheon (Tucson, AZ)
M393A3 and M467A1:
L-3 Communications (Lancaster, PA)

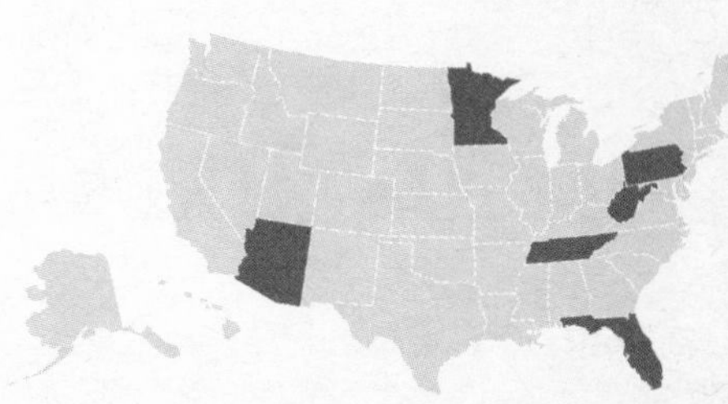

Thermal Weapon Sight II

Enables combat forces to acquire and engage targets with small arms, under day, night, obscurant, no-light, and adverse weather conditions.

INVESTMENT COMPONENT

- **Modernization**
- Recapitalization
- Maintenance

Description & Specifications

The AN/PAS-13 Thermal Weapon Sight (TWS) Generation II family enables individual and crew served gunners to see deep into the battlefield, increase situational awareness and target acquisition range, and penetrate obscurants, day or night. TWS II systems use forward looking infrared (FLIR) technology and provide a standard video output for training, image transfer or remote viewing. TWS II systems are silent, lightweight, compact, durable, battery powered thermal sight powered by commercial Lithium AA batteries. TWS II systems offer a minimum 20 percent longer range at roughly two-thirds the weight and with 50 percent power savings over the legacy TWS systems.

The TWS family comprises three variants:

- AN/PAS-13(V)1 Light Weapon Thermal Sight (LWTS) for the M16 and M4 series rifles and carbines as well as the M136 Light Anti-Armor Weapon.
- AN/PAS-13(V)2 Medium Weapon Thermal Sight (MWTS) for the M249 and M240B series medium machine guns.
- AN/PAS-13(V)3 Heavy Weapon Thermal Sight (HWTS) for the squad leaders weapon M16 and M4 series rifles and carbine, M24 and M107 Sniper Rifles, M2 HB and Mk19 machine guns.

Program Status

- **Current:** LTWS, MTWS, and HTWS Generation I systems in sustainment
- **Current:** TWS II in full-rate production
- **1QFY07:** Materiel release

Recent and Projected Activities

- **3QFY07:** Award follow on production contract
- **Ongoing:** Fielding and new equipment training

ACQUISITION PHASE

- Concept & Technology Development
- System Development & Demonstration
- **Production & Deployment**
- Operations & Support

Thermal Weapon Sight II

FOREIGN MILITARY SALES
None

CONTRACTORS
BAE Systems (Lexington, MA)
DRS (Melborne, FL; Dallas, TX; Irvine, CA)

Transportation Coordinators' Automated Information for Movement System II (TC-AIMS II)

Facilitates movements management of personnel, equipment, and supplies from home station to the conflict and back while providing source in-transit visibility data.

INVESTMENT COMPONENT

- **Modernization**
- Recapitalization
- Maintenance

Description & Specifications

Transportation Coordinators' Automated Information for Movements System II (TC-AIMS II) is a service migration system. Characteristics include:

- Source feeder system to Joint Force Requirements Generation II, Joint Planning and Execution System, Global Transportation Network, and Services' command and control systems
- Common user interface to facilitate multi-Service user training and operations
- Commercial off-the-shelf hardware/ software architecture
- Net-centric implementation with breakaway client-server and/or stand alone/workgroup configurations
- Incremental, block upgrade developmental strategy

Program Status

- **2QFY05–1QFY07:** Continue Block 2 fielding
- **2QFY05–1QFY07:** Complete development of Block 3, which will provide combatant commanders a reception, staging, onward movement, and integration capability, directly supporting in-theater transportation movement activities

Recent and Projected Activities

- **2QFY07–2QFY09:** Test of Block 3, with a milestone decision review anticipated in 4QFY07 to field Block 3

ACQUISITION PHASE

- **Concept & Technology Development**
- System Development & Demonstration
- Production & Deployment
- Operations & Support

Transportation Coordinators' Automated Information for Movement System II (TC-AIMS II)

FOREIGN MILITARY SALES
None

CONTRACTORS
Systems integration:
Computer Sciences Corp. (Falls Church, VA)
Program support:
L-3 Communications Titan Group (Newington, VA)
Facilities management:
Northrop Grumman (Alexandria, VA)

Tube-Launched, Optically-Tracked, Wire-Guided (TOW) Missiles

Provides long-range, heavy anti-tank and precision assault fire capabilities to the Army and Marine forces.

INVESTMENT COMPONENT

Modernization

Recapitalization

Maintenance

Description & Specifications

TOW (Tube-Launched, Optically-Tracked, Wire-Guided) is a heavy anti-tank/ precision assault weapon system, consisting of a launcher and a missile. The missile is 6 inches in diameter (encased, 8.6 inches), and 49 inches long. The gunner defines the aim point by maintaining the sight cross hairs on the target. The launcher automatically steers the missile along the line-of-sight toward the aim point via a pair of control wires, which physically link the missile and the launcher. The missile impact is at the aim point.

TOW missiles are employed on the High Mobility Multipurpose Wheeled Vehicle (HMMWV)-mounted Improved Target Acquisition System (ITAS), HMMWV-mounted M220A4 launcher (TOW 2), Stryker Anti-Tank Guided Missile Vehicles, and Bradley Fighting Vehicle Systems (A2/A2ODS/A2OIF/A3). TOW missiles are also employed on the Marine HMMWV-mounted M220A4 launcher (TOW 2), LAV-ATGM Vehicle, and AH1W Cobra attack helicopter. TOW is also employed by allied nations from a variety of ground and airborne platforms.

The TOW 2B Aero is the most modern and capable missile in the TOW family with an extended maximum range to 4,500 meters. This is accomplished with an increase of control wire and by affixing an aerodynamic nose to the missile. The TOW 2B Aero has an advanced counter active protection system capability. It defeats all current and projected threat armor systems. The TOW 2B Aero flies over the target (offset above the gunner's aim point) and uses a laser profilometer and magnetic sensor to detect and fire two downward-directed, explosively formed penetrator warheads into the target. The TOW 2B Aero's configuration weight is 49.8 pounds (encased, 65 pounds).

The TOW Bunker Buster is optimized for performance against urban structures, earthen bunkers, field fortifications, and light-skinned Armor threats. It has a 6.25 pound, 6-inch diameter high-explosive, bulk-charge warhead, and its missile weighs 45.2 pounds. The TOW BB has an impact sensor (crush switch) located in the main-charge ogive and a pyrotechnic detonation delay to enhance warhead effectiveness. The PBXN-109 explosive is housed in a thick casing for maximum performance. The TOW BB can produce a 21-24 inch diameter hole in an 8-inch thick, double-reinforced concrete wall at a range of 65 to 3,750 meters.

Program Status

TOW 2B Aero and Bunker Buster (BB)

- **4QFY97:** Last U.S. TOW 2B missile produced
- **2QFY04:** TOW 2B Aero multi-year production contract awarded for FY 04-06
- **3QFY05:** Awarded contract option for TOW Bunker Buster production for Army & U.S. Marine Corp
- **4QFY06:** Awarded FY06-09 Multi-Year contract for TOW Missile production

Recent and Projected Activities

- Continue U.S. production of TOW 2B Aero and Bunker Buster

ACQUISITION PHASE

Concept & Technology Development | System Development & Demonstration | **Production & Deployment** | Operations & Support

Tube-Launched, Optically-Tracked, Wire-Guided (TOW) Missiles

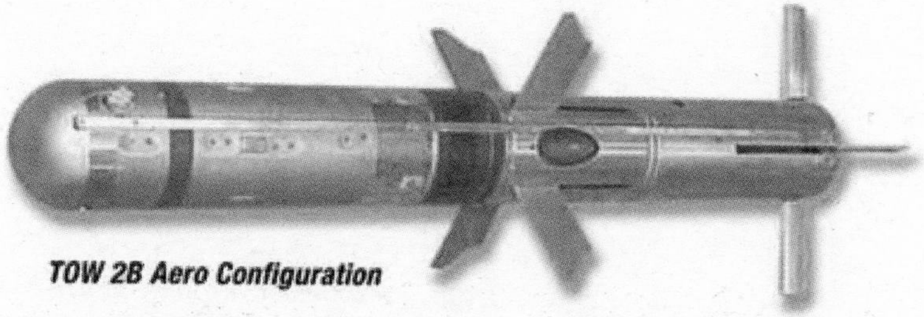

FOREIGN MILITARY SALES

The TOW weapon system has been sold through FMS to more than 43 allied nations over the life of the system

CONTRACTORS

TOW 2B Aero and TOW BB

Prime:
Raytheon (Tucson, AZ)

Control Actuator, Shutter Actuator:
Moog (Salt Lake City, UT)

Warheads:
Aerojet General (Socorro, NM)

Gyroscope:
Condor Pacific (Cheshire, CT)

Sensor:
Thales (Basingstoke, UK) (TOW 2B only)

Launch Motor:
ATK (Radford, VA)

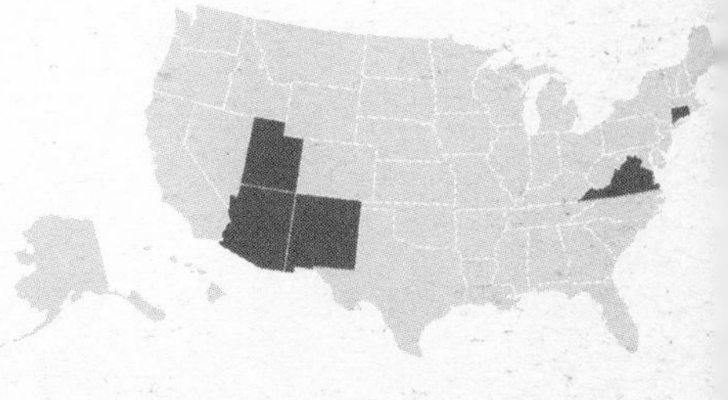

TOW 2B Aero Configuration

Unified Command Suite (UCS)

Provides voice and data communications to National Guard Weapons of Mass Destruction-Civil Support Team commanders to enhance assessment of and response to weapons of mass destruction events.

INVESTMENT COMPONENT

- Modernization
- Recapitalization
- Maintenance

Unified Command Suite deployed in New Orleans

Description & Specifications

The Unified Command Suite (UCS) vehicle is a self-contained, stand-alone C-130 air mobile communications platform intended to provide both voice and data communications capabilities to Civil Support Teams (CST) commanders.

The UCS consists of a combination of commercial off-the-shelf and existing government off-the-shelf communications equipment (both secure and non-secure data) to provide the full range of communications necessary to support the Civil Support Team mission. It is the primary means of reachback communications for the Analytical Laboratory System for the CSTs and acts as a command and control hub to deliver a common operational picture for planning and fulfilling an incident response. It provides:

- Digital voice and data over satellite network
- Non-Secure Internet Protocol Router Network (NIPRNET), Secure Internet Protocol Router Network (SIPRNET)
- Radio remote and intercom with cross-banding
- Over-the-horizon communication interoperable interface with state emergency management and other military units

Program Status

- **1QFY06:** System integration decision review
- **4QFY06:** Operational testing complete
- **1QFY07:** Integration & fielding IPR

Recent and Projected Activities

- **4QFY07:** Initial operational capability (IOC)
- **4QFY08:** Full operational capability (FOC)

ACQUISITION PHASE

- Concept & Technology Development
- System Development & Demonstration
- Production & Deployment
- Operations & Support

Unified Command Suite Deployed for Katrina Response

Unified Command Suite (UCS)

FOREIGN MILITARY SALES
None

CONTRACTORS
Vehicle:
Wolf Coach, Inc., an L-3 Company (Auburn, MA)
Communications system integrator:
Naval Air Warfare Center Aircraft Division (Patuxent River, MD)

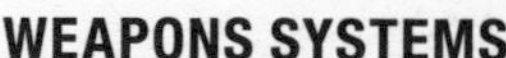

Unit Water Pod System (Camel)

Provides the Army with the capability to receive, store and dispense potable water to units at all echelons throughout the battlefield.

INVESTMENT COMPONENT

Modernization

Recapitalization

Maintenance

Description & Specifications

The Camel replaces the M107, M149 and M1112 series water trailers. It consists of an 800-900 gallon capacity baffled water tank, a thermal regulating module, and a filling stand for individual containers. It sits on a M1095 Trailer which allows for better transportability on and off the road by utilizing the Family of Medium Tactical Vehicle Truck. It holds a minimum of 800 gallons of water and provides 2+ days of supply at a minimum sustaining consumption rate. It is operational from -25 to +120 degrees Fahrenheit. The system also contains six filling positions for filling canteens and five gallon water cans. In the absence of dispensing water, the Camel can chill a full water payload ranging in temperature between 65 degrees Fahrenheit and 120 degrees Fahrenheit to 60 +/- 5 degrees Fahrenheit at a minimum rate of 1.5 degrees per hour. It also dispenses chilled water to 60 +/- 5 degrees Fahrenheit at the minimum rate of 40 gallons per hour.

Program Status

- **Current:** Production qualification testing (PQT) planned for January 07

Recent and Projected Activities

- **FY07:** Complete PQT, technical manuals
- **FY08:** Award Camel contract
- **FY09:** First unit equipped

ACQUISITION PHASE

Concept & Technology Development | System Development & Demonstration | Production & Deployment | Operations & Support

Unit Water Pod System (Camel)

FOREIGN MILITARY SALES
None

CONTRACTORS
Chenega Integrated Systems, LLC
(Panama City, FL)

Warfighter Information Network-Tactical (WIN-T)

Provides an integrating, high-speed, high-capacity backbone communications network that is focused on the Larger Operational Force.

INVESTMENT COMPONENT

- Modernization
- Recapitalization
- Maintenance

Description & Specifications

Warfighter Information Network-Tactical (WIN-T) is the Army's fully mobile, tactical communications system for reliable, secure, and seamless video, data, imagery, and voice services that enable decisive combat actions. It will move information in a manner that supports commanders, staff, and functional units while enabling the full mobility of large formations. WIN-T will establish an environment in which commanders at all echelons will have the ability to operate with virtual staffs and analytical centers that are located at remote locations throughout the battle space.

WIN-T is made up of separate configuration items consisting of switching, routing, and subscriber access nodes, transmission systems, and personal communications devices, fielded from Corps and above to battalion. This infrastructure guarantees high capacity throughput, assured security, and network management. It manages, prioritizes, and protects information in an uninterrupted means to get the right information to the right Soldier at the right time.

Army has shifted WIN-T's focus to support the Future Modular Force and Future Combat Systems. Its immediate priority is to enable the integration and testing of Points of Presence for Future Combat System platforms, with key delivery dates in 2007 – 2009. As select technologies become available, they will be introduced as potential capabilities to enhance Current Force systems such as the Joint Network Node. WIN-T will eventually completely replace the Joint Network Node network and a number of Current Force Systems as a single integrated network.

Program Status

- **1QFY07:** Operational requirements document converted to a capabilities development document

Recent and Projected Activities

- **2QFY07:** Preliminary design review (PDR) 1 with focus on FCS requirements
- **2QFY07–2QFY08:** Delivery of functional prototypes to Future Combat Systems
- **4QFY08:** PDR

ACQUISITION PHASE

- Concept & Technology Development
- System Development & Demonstration
- Production & Deployment
- Operations & Support

WIN-T 3-Tier Architecture

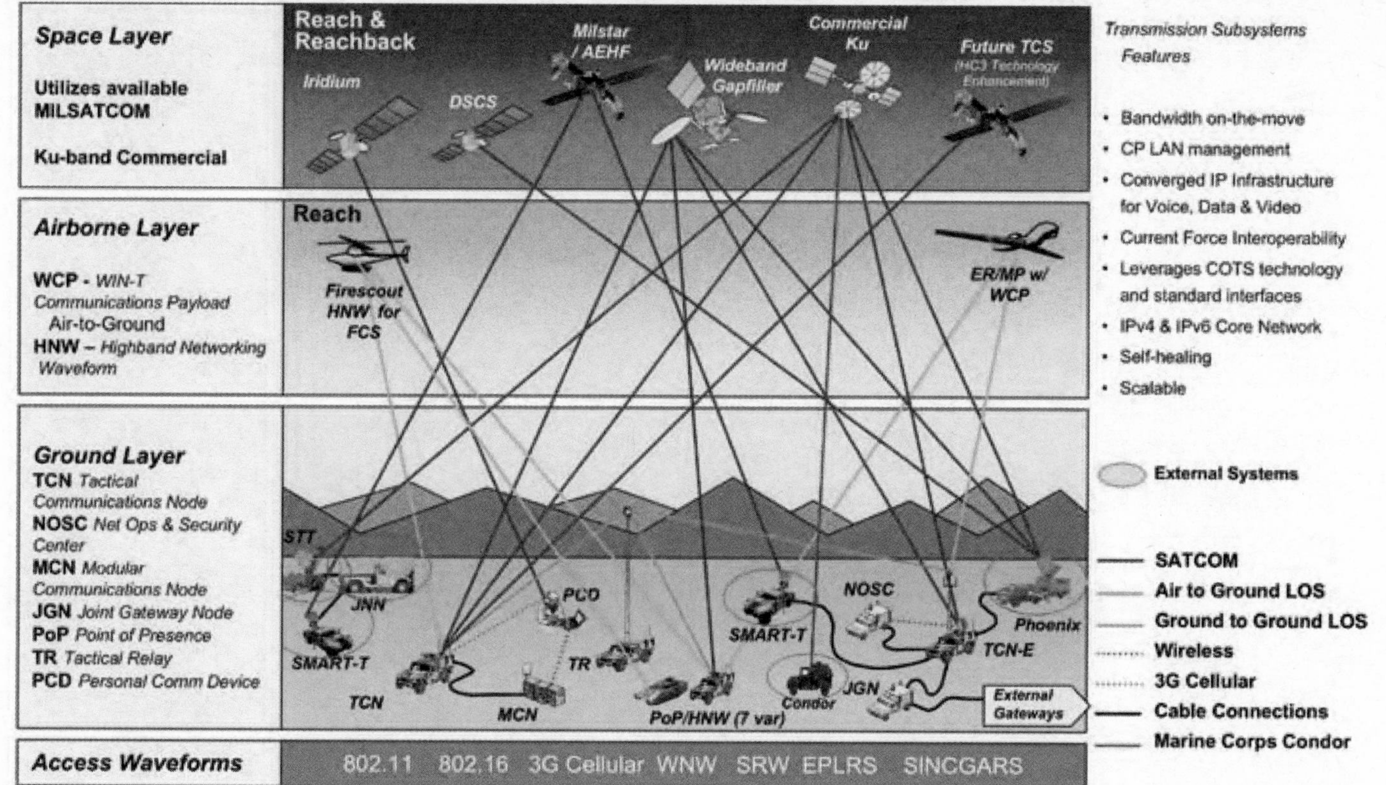

Warfighter Information Network-Tactical (WIN-T)

FOREIGN MILITARY SALES
None

CONTRACTORS
General Dynamics (Taunton, MA)
Lockheed Martin Mission Systems (Gaithersburg, MD)
Harris Corp. (Palm Bay, FL)
BAE Systems (Wayne, NJ)

XM101 Common Remotely Operated Weapon Station (CROWS)

Protects the gunner inside various lightly and heavily armored vehicles, including the up-armored High Mobility Multipurpose Wheeled Vehicle, while providing mobile, first-burst engagement of targets day or night.

INVESTMENT COMPONENT

Modernization

Recapitalization

Maintenance

Description & Specifications

The XM101 Common Remotely Operated Weapon System (CROWS) consists of a weapon mount, flat panel display, and a joystick controller. Within the mount are a day camera, thermal camera, laser range finder, meteorological sensors, and fiber optic gyroscopes. CROWS uses input from these sensors to calculate a ballistic solution to a target seen on the flat panel display. The Soldier uses the joystick controller to operate CROWS.

- Allows Soldiers to engage targets with current weapons while under armor
- Supported weapons: MK19, M2, M240B, and M249
- Day/night target engagement
- Two axis stabilized mount allows firing on the move
- Target track is independent of gun elevation movement

Program Status

- Fielding in support of urgent materiel releases
- **1QFY07:** Production verification testing

Recent and Projected Activities

- **1Q–3QFY07:** Long-term Army program under full and open competition
- **3QFY07:** Contract award

ACQUISITION PHASE

Concept & Technology Development | System Development & Demonstration | Production & Deployment | Operations & Support

XM101 Common Remotely Operated Weapon Station (CROWS)

FOREIGN MILITARY SALES
None

CONTRACTORS
Recon Optical, Inc. (Barrington, IL)
MICOR Industries, Inc. (Decatur, AL)

XM307

Provides vehicle and weapon squads with decisive overmatch capability against personnel and light armored targets with high explosive airbursting and armor-piercing ammunition.

INVESTMENT COMPONENT

- Modernization
- Recapitalization
- Maintenance

Description & Specifications

The XM307 Advanced Crew Served Weapon is a modern 25mm machine gun that is very light and imparts very low recoil forces to mounting platforms. The XM307 family of ammunition consists of a high explosive airbursting (HEAB) munition, an armor piercing munition, and a target practice munition. The XM307 HEAB munition is capable of defeating not only exposed targets but also those in defilade (targets that have taken cover behind structures, terrain features, or vehicles). The XM307's armor-piercing munition defeats lightly armored vehicles using a shaped charge projectile. The weapon also fires the target practice munition and has the capability to launch future non-lethal ammunition. The XM307 is a close combat weapon system for Future Combat System (FCS) vehicles and is being considered for other Brigade Combat Teams (BCTs). There is no current requirement for a ground-mounted system.

Program Status

- **1QFY04:** System development and demonstration contract awarded
- **4QFY05:** Program restructured for mounted and remote operation only

Recent and Projected Activities

- **FY04–FY09:** Conduct system development and demonstration activities
- **2QFY08:** Deliver initial XM307 remotely operated variant prototypes to FCS for integration efforts.
- **2QFY09:** Milestone C Decision
- **FY12–FY13:** Low rate initial production
- First unit equipped in accordance with FCS fielding schedule

ACQUISITION PHASE

Concept & Technology Development | **System Development & Demonstration** | Production & Deployment | Operations & Support

XM307

FOREIGN MILITARY SALES
None

CONTRACTORS
Prime:
General Dynamics Armament and Technical Products (Charlotte, NC)
Subcontractors:
Raytheon (El Segundo, CA)
Action Manufacturing (Philadelphia, PA)
Intertek Laboratories (Sterling, NJ)
EDO Corp. (New York, NY)

High Explosive Air-Burst Cartridge

Precision Air-Bursting
PBXN5 High Explosive
Defeats PASGT Vest & Helmet
Controlled Fragmentation Warhead

Armor Piercing Cartridge

51mm RHA (Threshold)
51mm HHA (Goal)

Target Practice Cartridge

Trainer

XM1050 TP | *XM1049 AP* | *XM1019 HEAB* | *XM1051 TP-S*

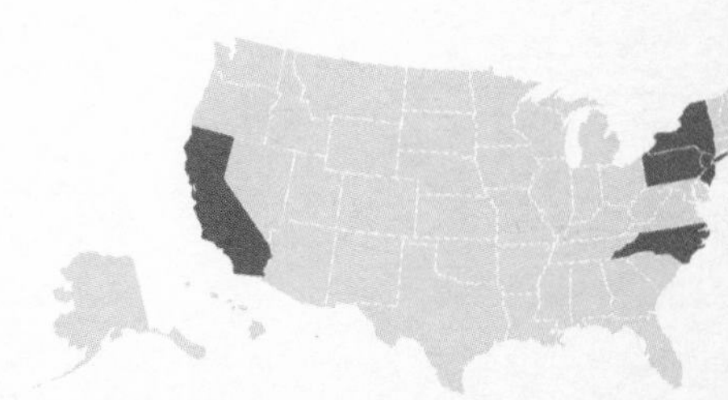

Science & Technology

The Army's Science and Technology (S&T) strategy is to develop the technology options that will enable the Future Force while seeking opportunities to enhance the Current Force. This strategy is achieved by simultaneously investing in the three components of Science and Technology:

1. **Research to create new understanding for paradigm shifting capabilities in the far-term,**
2. **Translating research into militarily useful technology applications in the mid-term, and**
3. **Demonstrating maturing technology in relevant operational environments and facilitating transfer of that technology during the near-term.**

Technology demonstrations "prove" the concept, inform the combat developments process, and provide the acquisition workforce with evidence of technology's readiness to satisfy system requirements. The diverse S&T investment portfolio exploits the dynamic nature of opportunities presented through scientific discovery and the "game-changing" potential of innovative technology applications in response to adaptive threat capabilities.

These investments are planned to achieve the capabilities outlined in the Quadrennial Defense Review (QDR) and the needs identified in the Training and Doctrine Command's (TRADOC) capability needs assessment. The Army program is synchronized with the DoD-wide S&T program through the director, Defense Research and Engineering Reliance 21 process.

S&T INVESTMENT—FUTURE FORCE TECHNOLOGY AREAS

The diverse Army S&T portfolio is characterized in terms of Future Force Technology Areas. The investments in these areas are shown opposite in a color depiction (Figure x-1) that approximates their proportionate dollar value in FY07 by Technology Area. The TRADOC documents depicted on the right describe the Army Capstone Concept, Future Capabilities Needs, and TRADOC Concept and Capabilities Development Plan (AC2DP). TRADOC conducts Capability Gap analyses annually to identify gaps and shortfalls to the S&T community that are used to shape Army Technology Objectives (ATOs) to satisfy specific needs within the gaps. The ATOs are the highest priority Army S&T efforts designated by the Headquarters of the Department of the Army (HQDA) funded within the Technology Area investments.

The S&T section of U.S. Army Weapons Systems is organized by Future Force Technology Area. Selected ATOs are described within most of the Technology Areas. ATOs are not designated within the Basic Research area because these investments fund sciences (discovery and understanding), not technology.

The ATOs are cosponsored by the S&T developer and the warfighter's representative, TRADOC. Capabilities derived from joint and Army concepts are used to identify and prioritize capabilities for the Future Force. TRADOC validates that the ATOs will provide enablers for needed capabilities. The ATOs are focused efforts that develop specific S&T products within the cost, schedule, and performance metrics assigned when they are approved. The goal is to mature technology within ATOs to transition to program managers for system development and demonstration and subsequently to acquisition. The complete portfolio of 86 ATOs is described in the 2007 Army Science and Technology Master Plan (distribution is limited to government and current government contractors).

Short descriptions of technologies or capabilities pursued within the Future Force Technology Area investments are provided below:

Force Protection technologies enable organizations, platforms, and Soldiers to avoid detection, acquisition, hit, penetration, and kill, including advanced armor, countermine and counter improvised explosive devices (IEDs) detection and neutralization, aircraft survivability, active protection systems, and installations.

Intelligence, Surveillance, and Reconnaissance (ISR) technologies enable persistent and integrated situational awareness and understanding to provide intelligence specific to the needs of the Soldier requirements, across the range of military operations.

Command, Control, Communications, and Computers (C4) technologies provide capabilities for superior decision-making, including intelligent network decision agents and antennas to link Soldiers, leaders, and organizations into a seamless battlefield network.

Lethality technologies enhance Soldiers and platforms to provide overmatch against threat capabilities and include non-lethal technologies enabling tailorable lethality options.

Medical research and technology protects and treats Soldiers to sustain combat strength, reduce casualties, and save lives.

Unmanned Systems technologies enhance the effectiveness of unmanned air and ground systems through improved perception, cooperative behaviors, and increased autonomy.

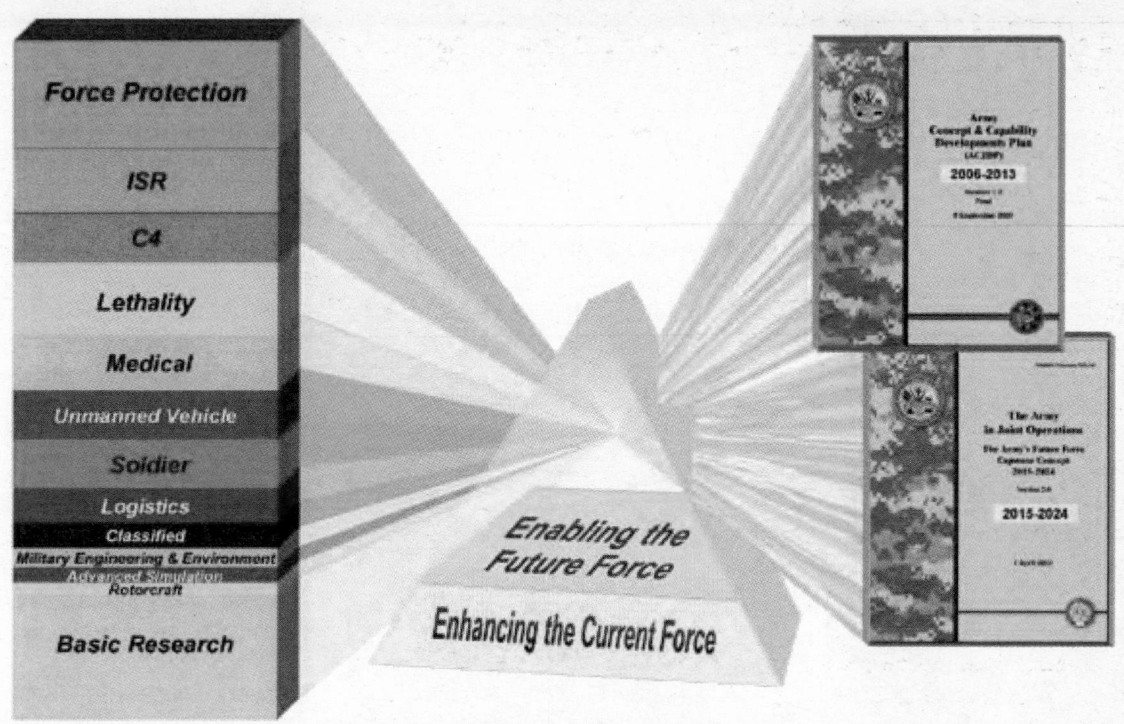

Figure x-1: The Technology Area color bands shown on the left are approximately proportional to the Army's financial investment in FY07. The specific technologies funded in these investment areas are aligned to achieve the Force Operating Capabilities defined by U.S. Army Training and Doctrine Command (TRADOC). The documents on the right identify the capability needs used in the TRADOC Capability Needs Analysis, which determine the Future Force Capability Gaps.

Soldier Systems technologies provide materiel solutions that protect, sustain, and equip Soldiers and non-materiel solutions that enhance human performance. Together these solutions enable Soldiers to adapt and excel against any threat.

Logistics technologies enhance strategic response and reduce logistics demand.

Military Engineering and Environment technologies enhance deployability, sustainability, and battlespace awareness. They also enable sustainment of training and testing range activities.

Advanced Simulation technologies provide increasingly realistic training and mission rehearsal environments to support battlefield operations, system acquisition, and requirements development.

Rotorcraft technologies enhance the performance and effectiveness of future rotorcraft.

Basic Research investments seek to develop new understanding to enable revolutionary advances or paradigm shifts in future operational capabilities.

The Army S&T program pursues technologies to enable a fully capable Future Combat Systems (FCS) Brigade Combat Team within the joint land force and to spin out technologies as they are available for the Current Force. The FCS related S&T investments are included within the Future Force Technology Area investments already described.

FORCE PROTECTION

Kinetic Energy Active Protection System (KEAPS)

The Kinetic Energy Active Protection System (KEAPS) ATO provides added capability to defeat tank-fired Kinetic Energy (KE) rounds to the Chemical Energy (CE) system that currently defines the FCS Point of Departure APS. This ATO develops warhead and interceptor chassis designs and will conduct robust component testing. These components support the FCS Hit Avoidance Suite designed to enhance the protection of FCS against tank-fired threats.

Mine and Improvised Explosive Device (IED) Detection

The Mine and IED Detection ATO addresses near-term technology advancements to defeat Current and Future Force mine/IED threats. This ATO demonstrates an increase in on-road speed for mounted forces' detection of anti-tank (AT) mines with low false alarm rates and capabilities for off-road detection of mines and roadside IEDs. It also demonstrates an autonomous, high-performance, standoff anti-personnel (AP) mine detection capability on a small unmanned ground vehicle (UGV) for increased survivability during dismounted mine/IED detection missions. This ATO develops new airborne sensor modalities to assist with mine/IED detection. This ATO also develops new techniques to automate change detection and speed up the image analysis process for IED detection. This effort completes development and integration of vehicular Ground Penetrating Radar (GPR); conducts a series of on-road and off-road vehicular GPR demonstrations in a variety of operational scenarios; fully integrates the cueing and confirmation sensor suite for roadside IED detection; and demonstrates a convoy escort capability.

Network Electronic Warfare/IED Countermeasures

This ATO delivers technology upgrades along with improved operational capabilities at the end of each FY, starting in FY06 to PM Counter RCIED Electronic Warfare (CREW). Each year will then see a progression of upgraded operational capabilities, by coupling previous efforts with enhancements and culminating with yearly system demonstrations.

Network Electronic Warfare will demonstrate UAV and ground-based electronic support measures systems in an operational environment. Information Operations algorithm development will focus on a broad range of target transmission parameters. This integrated capability may become the baseline for Spiral 4 of the Counter RCIED Electronic Warfare (CREW) program and spiral integration into FCS at TRL 6.

Countermine/IED Neutralization

The Countermine/IED Neutralization ATO enhances the operational tempo of the Future Force's mounted operations in maneuver areas, including urban operations, which have a high likelihood of surface and buried anti-tank mines and roadside IEDs.

INTELLIGENCE, SURVEILLANCE, RECONNAISSANCE

Soft Target Exploitation and Fusion (STEF)

The Soft Target Exploitation and Fusion (STEF) ATO develops and demonstrates automated tools to solve the fusion, exploitation, and sensor management/cross-cueing issues associated with prosecuting and tracking individuals, recognizing their patterns of association, and tracking the organizations they form. This effort allows the commander to target significant individuals and to understand the organizations exerting influence in his area of operation sufficiently to disrupt or attack the organizational infrastructure. STEF will develop a service-oriented architecture compliant framework and evaluate/develop basic fusion tools for relationship exploitation, sensor management, and cross-cueing. As the automated tools are matured, they will be spiraled into the DCGS-A program.

Human Infrastructure Detection and Exploitation (HIDE)

This effort focuses on the development of algorithms for the detection of human infrastructure presence, such as machinery, currents in wires, computer emanations, industrial compounds, and humans themselves in confined spaces, using a variety of low-cost sensors including acoustic, seismic, magnetic, electric-field, passive infrared (IR), chemical, Radio Frequency (RF), and optical imagers. Algorithms will be structured to be adaptable to varying combinations of sensor modalities, environmental conditions, and varying missions. The algorithms and sensors will be integrated on a small mobile Unmanned Ground Vehicle (UGV). Initial efforts will concentrate on a limited application for detection of man-made machinery and human activity in hidden/confined spaces. In the latter part of the ATO, coordination with CERDEC I2WD will tailor efforts for transition to meet program requirements. In addition, ARL is coordinating this ATO with other programs, such as Suite of Sense Through The Wall (STTW) Systems ATO, to ensure compatibility and avoid duplication of effort.

Multi-Mission Radar (MMR)

The Multi-Mission Radar (MMR) ATO develops a High Mobility Multipurpose Wheeled Vehicle/Brigade Combat Team-configured sensor for reconnaissance surveillance and target acquisition, situation awareness, alerting and cueing, and fire control quality information for air and missile defense (AMD) engagements.

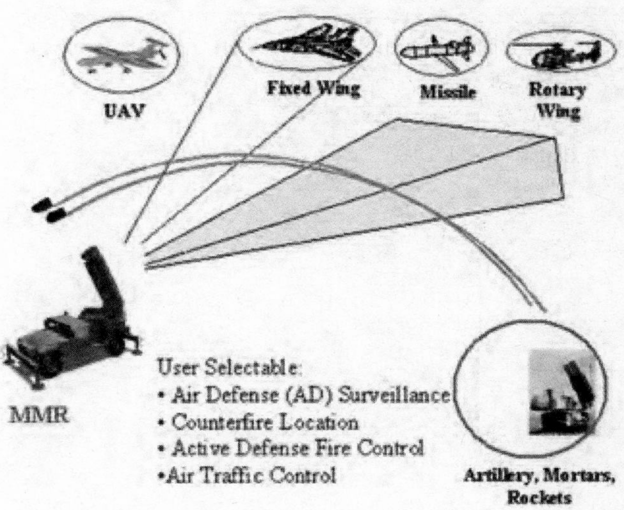

MMR will enable the Army to rapidly deploy a single sensor to perform multiple missions (e.g., AMD engagements of rockets, artillery, mortars, UAVs, cruise missiles, and rotary-and fixed-wing aircraft). Counter-Fire Target Acquisition will enable precision attack munitions while simultaneously providing data to maneuver and maneuver sustainment units. MMR with a towed generator will be small and light enough for insertion via single CH-47 or C-130 sortie, but will still have long-range target acquisition capabilities. This system will improve light force lethality and heavy force mobility and deployability. In addition, MMR will track friendly UAVs and aircraft, providing airspace management support for deconfliction and air traffic control. The effort will culminate in demonstrations of an MMR system and prime item development specifications suitable for moving into a system development and demonstration phase.

Suite of Sense Through the Wall Systems (STTW)

The Suite of Sense Through the Wall (STTW) systems will explore several technologies to provide mounted/dismounted users the capability to detect, locate, and "see" personnel with concealed weapons and explosives hidden behind walls, doors, and other obstructions. One version will be Soldier borne, and will be modularly designed to facilitate integration into FCS small unmanned

ground vehicles (SUGVs). Another will be UGV mounted and will have increased standoff distance from the target area. STTW will conduct limited evaluation of concealed explosives detection/concealed weapons detection and begin development of next generation STTW systems with limited standoff, improved target geolocation, and detection of multiple targets through walls. The ATO will work closely with emerging FCS and Future Force Warrior (FFW) network communications architectures to demonstrate transmission of STTW data on a real-time basis, and conduct lab testing of STTW prototypes, and user testing at Ft. Benning's Military Operations on Urban Terrain (MOUT) facility. Through experiments, STTW will develop tactics, techniques, and procedures and will characterize urban and complex terrain phenomenology. STTW will participate in both the Air Assault Expeditionary Force demo and the FFW Advanced Technology Demonstration (ATD) with handheld/Soldier borne STTW technology demonstrators.

Class II UAV Electro-Optical Payloads

The Class II UAV Electro-Optical Payloads ATO develops, integrates, and demonstrates a mission equipment package (MEP) to satisfy the reconnaissance, surveillance, and target acquisition (RSTA) requirements for the Class II UAV. Parallel development of technologies is executed in this program. (1) Line of sight (LOS) RSTA+laser designator integrates advanced compact infrared imaging sensors and a lightweight laser designator in an inertially stabilized gimbal. (2) Non-imaging LOS/through foliage RSTA incorporates laser vibrometry, acoustic, and/or magnetic anomaly-based sensor payloads. (3) Active imaging LOS/through foliage RSTA uses 2-D and 3-D LADAR techniques based on scanned/staring high bandwidth focal plane arrays, next-generation laser sources, optical modulation/mixing, and Geiger-mode processing. (4) Lightweight, efficient laser designator concepts that include pumping with high brightness laser diode arrays, passive Q-switching for simplicity, and alternative host materials that provide compact 50-100 milli-Joule designators are investigated for their utility in small UAV/UGV and soldier systems. The MEP concept provides target ID by enabling the scout to quickly see and characterize potential targets and non-target objects that are in the open and in complex/urban terrain obscured by modest foliage, camouflage, or other man-made or natural materials.

All-Terrain Radar for Tactical Exploitation of MTI and Imaging Surveillance (ARTEMIS)

The ARTEMIS ATO provides an all-weather/all-terrain airborne ground moving target indication, tracking, and cueing system for Class IV unmanned air vehicles (UAVs). Unlike most tactical radars, ARTEMIS will be able to track both mounted and dismounted threats that employ cover to conceal their movements or move in open terrain. ARTEMIS will also incorporate a synthetic aperture radar (SAR) capability sufficient to image vehicle-sized threats in foliated/open terrain and smaller threats that are in the open or shallowly buried. The program emphasizes signal and data processing achieving a prototype system early enough in the program to ensure the availability of real data for development. Tower testing will support risk reduction and acquire data needed for the development of signal processing algorithms. A decision point will occur at the end of tower testing to determine readiness for engineering flight tests to begin. The system will be integrated onto a manned surrogate platform for flight testing. Tower and flight test data will be collected to support adaptive MTI processing, advanced motion compensation, and advanced exploitation evaluation and incremental improvement.

C4

Networked Enabled C2 (NEC2)

The Network Enabled Command and Control (NEC2) ATO develops, integrates and transitions technologies, products, and services that provide network centric command and control (C2) capabilities to the Current and Future Force. Transition of these products and services are specifically focused on current, transitional (Battle Command Migration Plan), and future (FCS) Battle Command (BC) systems throughout all phases of operations and environments. NEC2 will develop advanced C2 software and algorithms that tailor and manage the flow of BC information and C2 services across Current and Future Force systems and enable the commander and his staff to effectively use vast amounts of information horizontally and vertically across the theater of operations for decision and information superiority. Technology efforts under NEC2 focus on applications in complex and urban terrain and battle command planning, execution, and re-planning products for unmanned systems and sensors, as well as decision making tools that account for political, religious, and cultural factors and expand the commander's reach to other government and non-government experts. The Unmanned Systems Capstone Experiment will evaluate unmanned software services v2.0 and air-ground systems performance across tactical application scenarios; collect and process communications characterization data; and deliver refined unmanned software services (v3.0, SRL 6) to PM FCS.

Tactical Mobile Networks (TMN)

The Tactical Mobile Networks (TMN) ATO develops, matures, and demonstrates communications and networking technologies that optimize bandwidth use, size, energy, and network prediction of tactical voice and data networks. TMN addresses emerging Future Force requirements through: (1) Proactive Diverse Link Selection (PAD-LS) algorithms to optimize use of available communications links; (2) multi-band, multi-mode tactical voice and data network communications services for dismounted Soldier Battle Command through the development of a Joint Tactical Radio System (JTRS) Software Communications Architecture (SCA) v2.2 Soldier Radio Waveform (SRW); and (3) software tools to dynamically predict and visualize on-the-move communications network performance.

TMN will conduct a modeling and simulation effort to develop and perform initial implementation and controlled environment demonstration of link selection algorithms. SRW will be implemented in prototype software defined radio (SDR) and will validate voice and data communications performance in operationally relevant field environment for dismounted Soldier and manned ground vehicle systems. Network management tools will be matured to include support for increased waveforms, entities, processing speed, full network topology, and a network visualizer to include network statistics and user priorities.

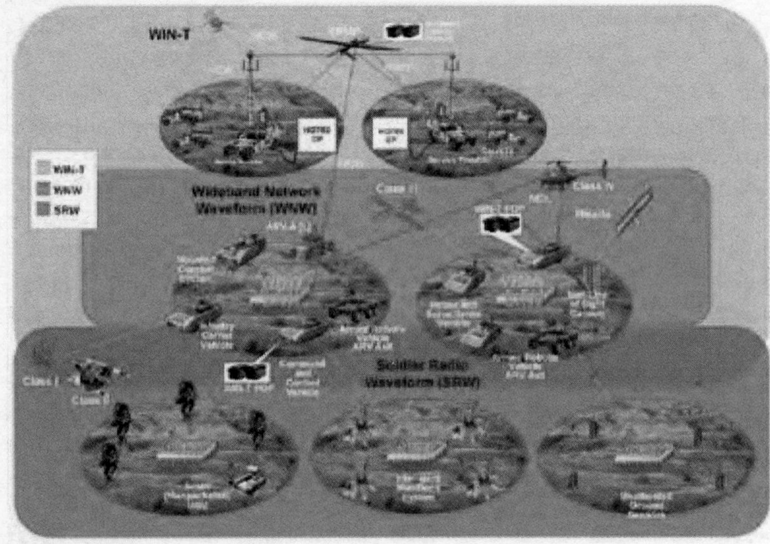

Tactical Network and Communications Antennas (TNCA)

The Tactical Network and Communications Antennas (TNCA) ATO develops, matures, and demonstrates affordable on-the-move (OTM) directional and omni-directional, low- profile antenna technologies and systems for the Current and Future Force. Efforts include affordable low-profile Ku/Ka and X band antenna systems and efficient Ku and Ka band power amplifiers for OTM tactical satellite communications (SATCOM). The antenna systems will be demonstrated in cooperation with PM WIN-T, JTRS JPEO, and FCS. Power amplifiers will leverage ARL's work with Gallium Nitride (GaN) and Metamorphic High Electron Mobility Transistor technologies. The effort will improve performance over current prototype multiband omni-directional antennas for OTM vehicular and dismounted Soldier platforms with ending TRLs of 6+. The omni antennas will provide higher gains to sustain Wideband Networking Waveform (WNW) link connectivity; reduce visual signatures; provide ballistic protection; host multiple waveforms on a single antenna; and provide for the integration of conformal, lightweight antennas within the Soldiers' protective combat gear, suppressing visual signatures and improving mobility. Body wearable antennas will be designed and developed to meet ground Soldier requirements.

Battle Space Terrain Reasoning and Awareness—Battle Command

This effort will provide integrated Battle Command capabilities to create and utilize actionable information from terrain, atmospheric, and weather effects and effects on units, systems, platforms, and Soldiers. This will enable agile, integrated ground and air operations in all operational environments. In FY10, an initial spiral of urban-based technologies from the Network Enabled Command and Control ATO will be incorporated. The resulting capability will result in Net Centric, n-Tier Terrain Reasoning Service(s) and embedded Battle Command applications.

This program will work with key transformational Battle Command programs and TRADOC Schools to: (1) conduct controlled demonstrations to gain insight into effectively integrating actionable terrain atmospheric and weather information into Battle Command SoS, staffs, processes and functions to enhance agile decision making and battle execution; (2) improve, extend, and mature terrain and weather-based information products and embedded applications within Battle Command SoS; (3) transition capabilities to DCGS-A, FCS, and CJMTK under forthcoming Technology Transition Agreements (TTA); and (4) support the development of a geo-Battle Management Language that extends the JC3IEDM to include representation of terrain, weather, and atmospheric actionable information.

LETHALITY

Non-Line of Sight-Launch System (NLOS-LS) Technology

The Non-Line of Sight-Launch System (NLOS-LS) Technology ATO is developing and maturing improved components and subsystem technologies for the NLOS-LS missile system. The ATO supports the NLOS-LS spiral development by transitioning affordable, mature components that enhance the NLOS-LS threshold performance through a subsystem maturation effort and continuing critical component development efforts for future NLOS-LS performance enhancements. By the end of FY07, this effort will perform component/subsystem bench testing, tower testing, and static/dynamic testing of critical component prototype hardware/software and begin captive flight test series for performance validation of enhanced seekers for the Precision Attack Missile (PAM). Products include improved seekers for better resolution, IM controllable propulsion maturation, warhead subsystem integration and testing, and validated simulation models and performance studies. Modeling and simulation efforts include the implementation of the collaboration of information and simulation technologies through the linkage of physics based engineering models, hardware/software in the loop (HWIL/SWIL) designs, constructive analysis, and virtual prototype development and exercise. The ATO is developing an affordable NLOS-LS missile with advanced imagery to enhance target detections and battle damage assessment, and missile propulsion technologies that provide extended range.

ElectroMagnetic (EM) Gun Technology Maturation & Demonstration

The ElectroMagnetic (EM) Gun Technology Maturation & Demonstration ATO focuses on developing and demonstrating key EM gun subsystems at or near full-scale to support future armament system developments. Future armored combat systems require more lethal yet compact main armament systems capable of defeating threat armor providing protection levels greatly in excess of current systems. The goal is to reduce technical risk associated with EM Gun technology by demonstrating meaningful technical progress at subsystem level; gain an understanding of EM technology issues; identify technology trends; conduct return on investment analyses; and craft a technology development strategy. By FY08, this effort will build a lightweight cantilevered high-fidelity railgun with integrated breech and muzzle shunt and demonstrate performance at hypervelocity and multi-round launch capability. It will integrate compact, twin counter-rotating pulsed alternator power supplies, conduct subsystem functional tests, and accomplish high fidelity PPS demonstrations that will establish requisite performance criteria to transition into the follow-on ATD. EM armaments offer the potential to field a leap-ahead capability by providing adjustable velocities, including hypervelocity, greatly above the ability of the conventional cannon. EM armaments could greatly reduce the sustainment requirements and vulnerabilities of conventional cannon systems and potentially can be fully integrated with electric propulsion and electromagnetic armor systems to provide an efficient, highly mobile, and deployable armored force. If successful, the payoff of EM gun technology will be increased lethality and lethality growth potential and enhanced platform survivability by reducing launch signature, and carrying less explosive energy on board.

Common Smart Submunition (CSS)

The Common Smart Submunition (CSS) ATO will develop and demonstrate the next generation target discriminating submunition. CSS will provide the warfighter with an affordable, smaller, one shot and more than one kill capability for gun and missile/UAV launch environments. The enhanced capabilities will be obtained through the integration of CSS's long standoff Explosively Formed Penetrator (EFP) warhead, laser radar (LADAR), and imaging infrared (IIR) sensor technologies into a small (120mm diameter) and lightweight submunition that can be deployed from multiple platforms (e.g., GMLRS, 155mm projectiles, 120mm mortars, UAVs). LADAR and IIR sensors will provide significant improved aimpoint correlation and discrimination capabilities. Substantial cost and logistics savings will be achieved through reduced munition size, multiple platform applications, and cross service use (Army, Air Force, Navy/USMC). In FY08, a full function submunition will undergo a captive carry test (CCT) to determine sensor performance against dynamic targets in a simulated tactical environment. Prototype submunitions demonstrating full form capability will be developed and drop tested. M&S system/analytical tools are being developed for platform evaluations and carrier integration analyses. The increased performance of CSS over the current submunition will result in a reduced logistics footprint and a significant decrease in unexploded submunitions.

Missile Seeker Technology

The Missile Seeker Technology ATO is developing advanced technologies for low-cost seekers and counter-measures for tactical missiles and future sensor applications. The ATO consolidates state-of-the-art research and development of advanced technologies in the areas of uncooled infrared (UCIR) technology for missile seekers and unmanned sensors, IR seeker counter-counter measures (CCM) for laser threats, and phased arrays for tactical seekers (PATS). This ATO will design, develop, and test advanced optics and signal processing techniques used in UCIR seeker and sensor packages; develop IR seeker CCM technologies to defeat near-term, Level II (Dazing) laser IR countermeasure (IRCM) threats and far-term, Level III (Damaging) laser IRCM threats; and develop Micro Electro-Mechanical Systems (MEMS)-based or alternative low-cost phased arrays to provide rapid beam steering for sensors, optical and RF missile seekers, and RF data links. This effort will conduct laboratory and field evaluation of lock-on-before-launch (LOBL) and lock-on-after-launch (LOAL) UCIR concepts, demonstrate prototype configurations, and transfer UCIR technology to appropriate aviation and missile systems. It will fabricate a passive phased sub-array with phase shifters, bench test, and initiate transition of PATS. The ATO will lead to lower cost, lighter weight, smaller weapons with increased lethality and reliability and enable the weapon systems to maintain lethality in the laser CM environment of the future battlefield.

Insensitive Munitions (IM) Technology

The Insensitive Munitions (IM) technology ATO is exploring ways to increase safety of items containing energetic materials. Increasing insensitivity of energetic materials will enhance survivability and sustainment of warfighting ability, while providing life-cycle cost savings for Army systems. Increased performance requirements (range and lethality) combined with lightening the force make reduced sensitivity munitions critical to the warfighter. Currently available technology does not enable IM-compliant munitions to be developed and fielded in timeframes consistent with requirements. New technologies, both energetics and system level mitigation, will provide solutions for designing munitions to maintain/improve survivability at reduced/constant platform and packaging weight, and obtain cost and logistics benefits through reduced hazard class and improved safety. This will provide a capability to PMs and PEOs to improve munitions response to IM threats for new or existing munitions. The ATO will explore new energetic-formulation and venting technologies for IM design, demonstrate integrated technologies for improved IM behavior in guns, missiles, and warheads, and develop predictive methodologies for IM tests. The payoff is improved tactical and combat system survivability leading to reduced transportation and storage burden.

MEDICAL

Automated Critical Care Life Support System (ACCLS)

The Automated Critical Care Life Support System (ACCLS) provides critically injured battlefield casualties with a self-contained life support system for far forward stabilization and transport in air and ground vehicles. This ATO will conduct research to support the discovery, adaptation, and development of algorithms and sensors for integration into a lightweight, semi-autonomous

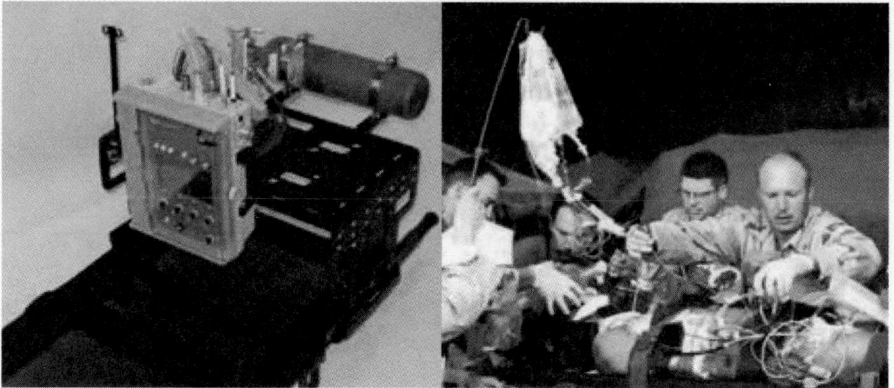

(Phase 1) and fully autonomous (Phase 2) critical care life support system, capable of monitoring and delivering closed-loop life support to combat casualties during treatment and within air and ground evacuation environments, including the UH60 and FCS Medical Treatment Vehicle variant. This ATO will produce clinically validated software algorithms for closed loop control of IV fluid and oxygen and will develop the system requirements and specification documents to generate a Request for Proposal (RFP) for the required hardware. The payoff of this approach is that treatment will be automatically titrated (i.e., adding small amounts until you reach desired medical effect) optimizing the casualty's condition while conserving fluid, oxygen, and power resources.

Fluid Resuscitation Tech to Reduce Injury and Loss of Life on the Battlefield

This ATO addresses noncompressible (i.e., any place a tourniquet or pressure cannot be applied) hemorrhage, which is the leading cause of death in casualties with potentially survivable injuries. Care of these patients requires control of bleeding and metabolic resuscitation. The solutions currently used for resuscitation, including blood products, dilute coagulation factors, and increase the tendency for more bleeding and metabolic imbalance. By integrating products formulated in other efforts and techniques to stop bleeding, metabolic resuscitation will be optimized and evaluated in relevant animal models. Clinical guidelines for the care of the combat casualty consisting of a mix of blood components, fluids, and drugs for the control of bleeding, and immunological stability of casualties, will be developed and evaluated.

Vaccines and Drugs to Prevent and Treat Malaria

This ATO aims to develop safe, effective vaccines to protect Soldiers against malaria and to develop drugs for the treatment of life-threatening malaria. Historically malaria is the disease that has caused the most disrupton in military operations in tropical regions of the world. There is no vaccine available for prevention, and the malaria parasite continues to develop resistance to new drugs used for treatment or prevention. Products from this ATO include vaccines for Plasmodium falciparum and Plasmodium vivax malaria, and a new FDA-approved drug for treating life-threatening severe and complicated malaria. This will result in improved protection against acquiring malaria and more effective and safer treatments if malaria is contracted.

Biomedical Enablers of Operational Health and Performance

This ATO provides multi-focused, biomedical solutions to protect Soldier health and enhance Soldier performance in extreme environments and during continuous operations. It will produce an altitude readiness management system that will enable unit commanders and mission planners to minimize the risk of altitude related injuries among Soldiers operating in high altitudes. The system will include strategies for rapid altitude acclimatization, and guidelines for determining medical and other logistical requirements in high-altitude operations. This ATO will lead to the development of new water doctrine for cold and mountain missions, as well as a mission planning tool that predicts average individual performance across a 0–48-hour period of sleep loss. Additionally, performance enhancing nutritional supplements for rations and predictive models based on high-altitude operations will be developed. This ATO will lead to the development of an enhanced fluid and nutrient delivery system to sustain hydration in Soldiers. This system will produce prediction models of warfighter water requirements; develop strategies, countermeasures, and doctrine to reduce hydration related injury; and sustain Soldier performance. Finally, the ATO will develop a mission planning tool for the prediction of individual performance during deployments. The payoff of this research will be improved health and performance in hot environments, high operational tempo, and at high altitudes.

UNMANNED SYSTEMS

Robotics Collaboration (RC)

The Robotics Collaboration (RC) ATO will develop advanced models, metrics, and design guidelines for optimal mounted and dismounted Soldier-robotic performance, and employ this information to demonstrate the technology required for effective interaction with both air and ground unmanned systems. This ATO will mature and demonstrate scaleable user interfaces for multi-screen mounted crew-stations, single screen workstations, or Soldier-portable PDA-sized devices. The interface design will also provide for graceful degradation of the display system by reconfiguring controls and displays in the event of hardware failure and provide associated functionality upon discovery of available services. RC will also develop an Intelligent Systems Behavior Simulator (ISBS) to mature and refine mounted crew and dismounted Soldier task models and associated metrics. Results from iterative ISBS experimentation will drive development of intelligent agents that decrease Soldier workload and reduce and/or automate controlling tasks across mounted and dismounted systems. It will develop 3-D models and algorithms using colorized ranging with LADAR and visual sensors for safe operations of unmanned systems around Soldiers and pedestrians. Models will be tested in both man-in-the-loop simulation and field experimentation using FCS-relevant scenarios. Hardware and software will be integrated onto existing manned and unmanned platforms. Additionally, this ATO will develop and demonstrate militarily-relevant autonomous collaborative behaviors for multiple unmanned vehicles (UVs), simultaneously supporting both mounted/dismounted ground and airborne warfighters.

Near Autonomous Unmanned Systems

This effort will develop, integrate, and demonstrate robotic technologies required for Future Force unmanned systems.

This ATO will advance the state of the art in perception and control technologies to enable the unmanned platform to conduct missions autonomously in populated, dynamic, and complex environments while adapting to changing conditions. It will develop initial tactical/mission behavior technologies to provide the unmanned ground vehicle (UGV) the capability to maneuver tactically in conjunction with manned systems and provide the Soldier-machine interface technology that will permit Soldiers to control multiple types of air and ground unmanned systems. A UGV self-security system will be developed and integrated to allow the detection of human threats at varying distances from the vehicle and calculate the appropriate response to deter the threat. Modeling and simulation will be used to develop, test, and evaluate the unmanned systems technologies (i.e., tactical behaviors, UGV self-security, perception algorithms). A test bed platform and appropriate mission modules will be integrated with the software and associated hardware developed to support warfighter experiments in a militarily significant environment.

Manned-Unmanned Rotorcraft Enhanced Survivability (MURES)

Current threat warning receivers (radar, missile, laser) and situational awareness systems, such as Integrated Situational Awareness & Targeting (ISAT), provide threat identification and sector/geolocation only. Existing decision aiding systems, such as Rotorcraft Pilot's Associate (RPA), provide line-of-sight masking to threat sensors only. These systems lack the intelligence to provide the critical information own force assets really need. Technology exists to assemble threat lethality footprints in real time by assessing own ship signatures, clutter environment, threat sensitivity, and jammer effectiveness. This information is a critical element to effective team decision aiding and management of manned and unmanned aircraft operating in threat environments, and directly enhances team survivability and mission effectiveness.

This ATO will develop real-time Survivability Planner Associate Rerouter (SPAR) software tailored to small unit manned-unmanned team operations. It will also develop interfaces to existing unmanned and manned mission management/decision aiding architectures to allow team cooperative/collaborative responses to tactical threats and enhance team survivability. Specific demonstration goals are rapid (<1 sec) and accurate (>90%) situational understanding of threat lethality to own force assets through Future Army mission simulations.

SOLDIER SYSTEMS

Leader Adaptability

This ATO provides the strategies and methods for rapidly developing the skills that leaders will need to perform adaptively in current mission scenarios and in future net-enabled environments. Army leaders already face new and rapidly changing challenges in the operational environment. In the future they will be challenged with a more widely distributed battlefield. Future technologies will provide increasing amounts of information from a variety of sources. They will have to employ network systems and decision aids not currently available. Leaders at the tactical level must learn to adapt more quickly to unanticipated operational situations and to fully advantage new capabilities as they are fielded. Prior research demonstrated interactive leader training tools to develop agility, adaptability, and selected critical thinking skills. This research will be leveraged to develop and assess training and leader development methods to improve how the Army develops leadership and battle command skills, including procedures for near real-time performance assessment and feedback. The methods will support faster skill acquisition and longer retention of cognitive and interpersonal skills required for adaptable performance. In FY08, this ATO will demonstrate technologies to train leadership skills in junior leaders through synthetic experience and develop prototype training programs to improve learning of complex cognitive skills necessary for adaptive battle command performance in net-enabled environments.

Future Force Warrior

The Future Force Warrior (FFW) Advanced Technology Demonstration (ATD) is developing and demonstrating an integrated Soldier and Small Team System of Systems (SoS). The FFW ATD is achieving revolutionary advances for the dismounted warfighter by developing Networked Communications/Collaborative Situational Awareness (NC/CSA), an integrated modular combat ensemble, netted lethality, man-portable power sources, Soldier mobility/sustainability, and human performance monitoring for the Future Force. This FFW SoS will be network

compatible with the FCS, other Future Force platforms, and robotic air/ground platforms to form adaptive, distributed sensor networks for better warfighter situational awareness and the capability to leverage combat power from higher echelons. Key performance goals for the ATD include developing and integrating a networked Soldier system that weighs less than 70 lbs in the fighting load (rifleman configuration, night attack, respiratory CB threat), supports 24-hour individual and 72-hour autonomous team operations, supports full NC/CSA with networked fires, and has connectivity with the network through leveraging the SLICE Warrior System Radio Terminal (WSRT) hosting the Soldier Radio Waveform (SRW). The goal is to demonstrate 2X increased combat effectiveness of Soldiers in small combat units (SCU) and improve affordability of the ensemble in preparation for transition to Ground Soldier System (GSS) system development and demonstration. In addition the ATD will identify/transition mature technologies for insertion into Land Warrior and analyze the other variants (core, mounted, air) to identify common requirements and gaps to enable expansion of the FFW/GSS architecture and system design concept for the other variants. The FFW ATO is developing and integrating component hardware and software and building a limited number of prototype systems to conduct limited user evaluations. The FFW architecture, design, hardware and software prototypes, and user evaluations will support the GSS Milestone B decision.

Mounted/Dismounted Soldier Power

The Mounted/Dismounted Soldier Power ATO delivers critical power to the battlefield for essential C4ISR equipment required by the Future Force. This ATO will develop, demonstrate, and transition component power technologies leading to higher-energy, lighter-weight, quieter, fuel- and cost-efficient power sources, battery chargers, and cooling systems. It includes the development of Micro Electro-Mechanical Systems (MEMS) technology to further increase power system efficiency and reduce size and weight. This effort will provide technology advancements leading to new silent watch capability with quiet, lightweight mobile power generation. The ATO delivers 50% fuel savings with co-generation of cooling and power for shelter/tent systems. It bridges the gap between vehicles and Soldiers with stand-alone self-powered man-portable field chargers/remote power sources, which reduce logistics costs 80% by allowing the tactical use of rechargeable batteries. Specific power solutions and goals will include component level development of burner technologies; component development and integration of heat driven cooling system technologies combined with waste heat recovery systems; improved energy density and recharge rate in rechargeable batteries; component integration of meso-machine, stirling, and fuel cell systems; and MEMS coolers for advanced passive and active cooling at the device and module level for high-power electronics.

Soldier Protection Technologies

This ATO focuses on the development of innovative materials for lightweight ballistic/blast and predictive tools for the development of improved individual armor systems. This effort will research and develop novel fiber technologies to reduce the weight of ballistic protection for the individual warfighter, while maintaining or increasing ballistic protection from conventional and emerging threats. The effect of novel blast weapons on currently fielded protective clothing and its interaction with the human body will be researched to obtain a better understanding of the energy transfer mechanisms inherent in protective materials and within the body. This information, coupled with data on the interaction of blast waves with textile materials, will allow the development of physics-based models and analytical tools for the design of clothing systems that optimally attenuate the effects of blast overpressure, and an experimental test device for evaluating the designs. A product of the ATO is a software tool capable of identifying promising extremity armor configurations. This will guide the design of new extremity armor configurations that will provide individual Soldiers with ballistic protection that offers maximum flexibility, agility, and mobility and minimizes the energy expended during movement in dismounted operations.

LOGISTICS

Advanced Lightweight Track

The Advanced Lightweight Track program will develop a new, field supportable, robust track system with lightweight, low-vibration, and low-acoustic emissions as well as reduced crew maintenance. The advanced hybrid steel track concept will offer greater robustness with a nominal weight increase over the continuous segmented band track.

The development of a lightweight track system presents some challenges. For example, traditional lightweight track is less durable and more vulnerable to mine blast damage. To overcome these challenges, a new design approach and new elastomer components will be incorporated into the development of two new track families, (i.e., the segmented band track and the hybrid steel track). These approaches will yield a new generation of track systems with both the lightweight and robust capabilities for future tracked combat vehicles.

Precision Airdrop–Medium

The Precision Airdrop–Medium ATO is developing technologies for medium payload weight precision airdrop, a capability the Army currently lacks. This is part of the Joint Precision Airdrop System, a family of guided airdrop systems. This ATO is expanding upon the current precision airdrop weight capability of 10,000 lbs to a 30,000 lbs capacity, potentially providing high-volume resupply of fuel and ammo, along with other sustainment cargo. In addition, it will enhance the potential strategic deployment airdrop capability to augment air/land and mitigate the maximum on ground (MOG) limitations (high-volume supplies and combat equipment). The ATO is developing a decelerator/parachute system and guidance and navigational capabilities to deploy a 30,000 lbs payload from 25,000 ft., with an airdrop accuracy of 100m CEP. The payload weight aligns with the medium block for Precision, Extended Glide Airdrop System (PEGASYS) and the corresponding payload capacity of the Palletized Load System/Load Handling System (PLS/LHS) and the related, emerging technology program Smart Distribution-Modular Intermodal Platform (MIP). A medium weight precision offset airdrop capability will enable immediate tactical deployability, as well as reduced drop zone detectability and vulnerability. The ATO is developing the technology to provide the required never-too-late supply and distribution capability the widely dispersed combat teams of the Future Force will require in the first days of a conflict, increasing their operational agility.

Hybrid Electric FCS

The Hybrid Electric drive program matures and characterizes technologies for FCS ground vehicles to enable silent operation/mobility, enhanced dash speed, battlefield robustness, and reduced signatures (acoustic, thermal, visual, electromagnetic interference.) This ATO focuses on reducing weight/volume by increasing operating frequencies of the power electronic switching devices (inverters and dc-dc converters). It also focuses on reducing the thermal management size, weight, and power requirements by increasing the operating temperatures of the individual system components (batteries, converters/ inverters, motors/generators). This program advances the Hybrid Electric Vehicle (HEV) System Architecture, allowing for intelligent power and energy generation and management and includes the development of a hybrid electric test cycle baseline to measure the variables of hybrid vehicles against standard vehicles. The fuel economy tests will evaluate Future Tactical Truck System (FTTS) variants and other existing HEV assets against standard baseline vehicles.

Products of this ATO will be integrated into the power and energy hardware in the Loop Systems Integration Laboratory (SIL) for further evaluation to determine their impact on hybrid electric power system performance. Development and validation efforts are coordinated closely with the FCS LSI and Manned Ground Vehicle (MGV) integrators.

Prognostics and Diagnostics for Operational Readiness and Condition Based Maintenance

This ATO develops prognostics algorithms and application-specific sensors. The Common Logistical Operating Environment-compatible core sensors-processing module will interface these algorithms and application-sensors with low/no-power THVS (temperature, humidity, vibration, and true g-shock) sensors and processor via secure wireless link for remote interrogation. Near-term and FCS commodity readiness and maintainability rely on the ability to detect health status and performance and environmental conditions/metrics that limit asset lifetime. Diagnostic sensors will enable health assessment. Predicting remaining lifetime requires interpretation of the information on both the asset and quality of sensor data. Commanders and logisticians must be able to access the data expeditiously with a minimum of effort. This project will develop a core "tag" with embedded sensors and processing that can be wirelessly interrogated. The asset's sensor history data will be analyzed by both on-board and post-processed prognostics algorithms to assess immediate readiness and remaining time to maintenance or lifetime. Resultant data will yield actionable information for both commander and logistician leading to increased readiness, enhanced awareness of materiel condition, increased confidence of mission completion, and smaller logistics footprint through condition-based maintenance.

ADVANCED SIMULATION

Learning with Adaptive Simulation and Training (LAST)

Army trainers lack the ability to rapidly create effective, Common Operating Environment (COE)-relevant, virtual training simulations that incorporate political/cultural effects of the environment and behaviors of an adaptive, asymmetrical enemy. LAST focuses on developing the pedagogical design and enhanced tools/methods for rapidly creating/modifying COE-relevant scenarios in virtual simulations. Additionally, the research develops enhanced virtual entities and political/cultural effects in integrated virtual simulations. These efforts will produce improvements in instruction, decision-making, and learning retention. The research leverages the abilities of the RDECOM STTC, ARI, ARL-HRED, and ICT. We have developed a scenario task list based on cognitive task assessment, a pedagogical design for the user environment, and a concept for the single user module practice environment for bilateral engagements. We have identified an initial set of tools/methods and behavioral models and are integrating these tools/

methods with a single-user module providing training in bilateral negotiation. The effort will conclude by conducting warfighter experiments and assessing the effectiveness of prototypes.

Scaleable Embedded Training and Mission Rehearsal (SET-MR)

Embedded training (ET) is a key performance parameter for FCS and GSS, and is required by Abrams, Bradley, and Stryker platforms, but it has been slow to evolve. FCS is developing live, virtual, and constructive (LVC) ET but has no significant ET spinout to the Current Force until FY12 and Current Force ET fielding plans are limited to gunnery training. This ATO will support a common implementation strategy and address known technology shortfalls in ET across Current Force combat and GSS. It aims to accelerate ET and mission rehearsal (MR) implementation in Current Force combat vehicles and Soldier systems up to platoon level. It will also develop tactical engagement simulation (TES) sensors for dismounted Soldier live training, size, power, and accuracy requirements. Additionally this ATO will provide ET risk mitigation for FCS, GSS, Heavy Brigade Combat Team (HBCT), and Stryker Brigade Combat Team (SBCT). The ATO will end in FY09 with field demonstrations of mission rehearsal and LVC ET using Current Force combat vehicles and dismounted Soldiers as the experimental force.

S&T ROLE IN FORMAL ACQUISITION MILESTONES

The Army S&T community role in acquisition involves not only technology development and transition but also formal participation in Milestone Decisions for acquisition programs of record. At Milestone A, the DASA (R&T) needs to ensure that the Technology Development Strategy for a program is synchronized with the Army S&T program and that the Technology Transition plan is realistic and funded. As the component S&T Executive, the DASA (R&T) is responsible for conducting a Technology Readiness Assessment (TRA) at all Milestone B and Milestone C decisions for Major Defense Acquisition Programs (MDAP). This assessment has become even more important with recent statutory requirements for the Milestone Decision Authority (MDA) to certify to Congress that the technologies of an MDAP have been demonstrated in a relevant environment–prior to approving a Milestone B. The TRA serves as the gauge of this readiness for the MDA's certification at both Army and Office of the Secretary of Defense (OSD) levels. The TRA process is a collaborative effort carried out among the Program Office, the S&T community, and (for ACAT 1D programs) the Under Secretary of Defense for Acquisition, Technology, and Logistics.

SUMMARY

The Army Science and Technology Investments seek to enable capabilities described in the Quadrennial Defense Review and the needs established in the TRADOC Capability Gap/Technology Shortfalls process. The S&T investments are characterized in Future Force technology areas. The highest priority S&T investments in Future Force technology areas are designated as ATOs. The ATOs are developed and approved for execution through a rigorous process that engages the S&T, acquisition, and combat development communities. Each ATO has defined products, milestones, designated resources, and projected warfighter payoff. The ATO products are shaped to provide technologies that are relevant to satisfying capabilities needs. The acquisition program managers partner with the S&T materiel developers to enhance opportunities for rapid transition of technology described in Technology Transition Agreements.

Appendices

Army Combat Organizations

Army organizations are inherently built around people and the tasks they must perform. Major combat organizations are composed of smaller forces, as shown here.

Squad

- Leader is a sergeant
- Smallest unit in Army organization
- Size varies depending on type–Infantry (9 Soldiers), Armor (4 Soldiers), Engineer (10 Soldiers)
- Three or four squads make up a platoon

Platoon

- Leader is a lieutenant
- Size varies—Infantry (40 Soldiers), Armor (4 tanks, 16 Soldiers)
- Three or four platoons make up a company

Company

- Leader is a captain
- Usually up to 220 Soldiers
- Artillery unit of this size is called a battery
- Armored Cavalry or Air Cavalry unit is called a troop
- Basic tactical element of the maneuver battalion or cavalry squadron
- Normally five companies make up a battalion

Battalion

- Leader is a lieutenant colonel
- Tactically and administratively self-sufficient
- Armored Cavalry and Air Cavalry equivalents are called squadrons
- Two or more combat battalions make up a brigade

Brigade

- Leader is a colonel
- May be employed on independent or semi-independent operations
- Combat, combat support, or service support elements may be attached to perform specific missions
- Normally three combat brigades are in a division

Division

- Leader is a major general
- Fully structured division has own brigade-size artillery, aviation, engineer, combat support, and service elements
- Two or more divisions make up a corps commanded by a lieutenant general

Glossary of Terms

Acquisition Categories (ACAT)

ACAT I programs are Milestone Decision Authority Programs (MDAPs [see also Major Defense Acquisition Program]) or programs designated ACAT I by the Milestone Decision Authority (MDA [see also Milestone Decision Authority]). ACAT I programs have two sub-categories:

1. ACAT ID, for which the MDA is USD (A&T). The "D" refers to the Defense Acquisition Board (DAB), which advises the USD (A&T) at major decision points.
2. ACAT IC, for which the MDA is the DoD Component Head or, if delegated, the DoD Component Acquisition Executive (CAE). The "C" refers to Component. The USD (A&T) designates programs as ACAT ID or ACAT IC.

ACAT IA programs are MAISs (see also Major Automated Information System (MAIS) Acquisition Program), or programs designated by the Assistant Secretary of Defense for Command, Control, Communications, and Intelligence (ASD [C3I]) to be ACAT IA. A MAIS is an AIS acquisition program that is:

1. Designated by the ASD (C3I) as a MAIS, or
2. Estimated to require program costs in any single year in excess of $30 million in FY96 constant dollars, total program costs in excess of $120 million in FY96 constant dollars, or total life-cycle costs in excess of $360 million in FY96 constant dollars.

ACAT IA programs have two sub-categories:

1. **ACAT IAM**, for which the MDA is the Chief Information Officer (CIO) of the DoD, the ASD (C3I). The "M" (in ACAT IAM) refers to Major Automated Information System Review Council (MAISRC). (Change 4, 5000.2-R)
2. **ACAT IAC**, for which the DoD CIO has delegated milestone decision authority to the CAE or Component CIO. The "C" (in ACAT IAC) refers to Component.

ACAT II programs are defined as those acquisition programs that do not meet the criteria for an ACAT I program, but do meet the criteria for a major system, or are programs designated ACAT II by the MDA.

ACAT III programs are defined as those acquisition programs that do not meet the criteria for an ACAT I, an ACAT IA, or an ACAT II. The MDA is designated by the CAE and shall be at the lowest appropriate level. This category includes less-than-major AISs.

Acquisition Phase

All the tasks and activities needed to bring a program to the next major milestone occur during an acquisition phase. Phases provide a logical means of progressively translating broadly stated mission needs into well-defined system-specific requirements and ultimately into operationally effective, suitable, and survivable systems. The acquisition phases for the systems described in this handbook are defined below:

Concept and Technology Development

Concept and Technology Development refers to the development of a materiel solution to an identified, validated need. During this phase, the Mission Needs Statement (MNS) is approved, technology issues are considered, and possible alternatives are identified. In this phase, the initiation concept is approved, a lead Component is designated, and exit criteria are established. The leader of the concept development team will work with the integrated test team to develop an evaluation strategy that describes how the capabilities will be evaluated once the system is developed.

Major components of this phase are Concept Exploration, Decision Review, and Component Advanced Development. Concept Exploration evaluates the feasibility of alternative concepts and assesses the merits of these concepts. This phase ends with a Decision Review, at which the preferred concept for technologies available is selected. The Decision Review may also determine whether additional component development is necessary before key technologies can enter System Development and Demonstration. Component Advance Development occurs when the project leader has a concept for the needed capability, but does not yet know the system architecture. The project exits Component Advanced Development when a system architecture has been developed and the component technology has been demonstrated in the relevant environment or the Milestone Decision Authority (MDA) decides to end this effort. This effort is intended to reduce risk on components that have only been demonstrated in a laboratory environment and to determine the appropriate set of subsystems to be integrated into a full system.

System Development and Demonstration

System development and demonstration is the process of developing concepts into producible and deployable products that provide capability to the user. The purpose of this phase is to develop a system, reduce program risk, ensure operational supportability, design for producibility, ensure affordability, and demonstrate system integration, interoperability, and utility. The major components of this phase are System Integration, System Demonstration, and Interim Progress Review. Development is aided by the use of simulation-based acquisition and guided by a system acquisition strategy and test and evaluation master plan (TEMP). System modeling, simulation, and test and evaluation activities are integrated into an efficient continuum planned and executed by a test and evaluation integrated product team (T&E IPT).

The independent planning, execution, and evaluation of dedicated Initial Operation Test and Evaluation (IOT&E), as required by law, and Follow-on Operational Test and Evaluation (FOT&E), if required, are the responsibility of the appropriate operational test activity (OTA). The program enters System Integration when the Project Manager has an architecture for the system, but has not yet integrated the subsystems into a complete system. This effort is intended to integrate the subsystems and reduce system-level risk. The purpose of the Interim Progress Review is to confirm that the program is progressing as planned or to adjust the plan to better accommodate progress made to date, changed circumstances, or both. The program enters System Demonstration when the Project Manager has demonstrated the system in prototype articles.

Production and Deployment

The purpose of the Production and Deployment phase is to achieve an operational capability that satisfies mission needs. In this phase, software has to prove its maturity level prior to deploying to the operational environment. Once maturity has been proven the system or block is baselined and a methodical and synchronized deployment plan is implemented to all applicable locations. A system must be demonstrated before DoD will commit to production and deployment. For DOT&E Oversight programs, a system cannot be produced at full-rate until a Beyond Low Rate Initial Production Report has been completed and sent to Congress, the Secretary of Defense, and the USD (AT&L).

The components of this phase include Low Rate Initial Production (LRIP), the Full-Rate Production Decision Review, and Full-Rate Production and Deployment. LRIP is intended to result in completion of manufacturing development to ensure adequate manufacturing capability and to produce the minimum quantity necessary for initial operational test and evaluation. The Full-Rate Production Decision Review considers the cost estimate, manpower, results of test and evaluation, compliance and interoperability certification. Following the completion of a Full-Rate Production Decision Review, the program enters Full Rate Production and Deployment.

Operations and Support

The objective of the Operations and Support phase is the execution of a support program that meets operational support performance requirements and sustainment of systems in the most cost-effective manner throughout their life-cycle. The sustainment program includes all elements necessary to maintain the readiness and operational capability of deployed systems. The scope of support varies among programs but generally includes supply, maintenance, transportation, sustaining engineering, data management, configuration management, manpower, personnel, training, habitability, survivability, safety, IT supportability, and environmental management functions. This activity also includes the execution of operational support plans.

Programs with software components must be capable of responding to emerging requirements that will require software modification or periodic enhancements after a system is deployed. A follow-on operational test and evaluation program that evaluates operational effectiveness, survivability, suitability, and interoperability, and that identifies deficiencies is conducted, as appropriate.

Acquisition Program

A directed, funded effort designed to provide a new, improved or continuing weapons system or AIS capability in response to a validated operational need. Acquisition programs are divided into different categories that are established to facilitate decentralized decision-making, and execution and compliance with statutory requirements.

Advanced Concept Technology Demonstrations (ACTDs): ACTDs are a means of demonstrating the use of emerging or mature technology to address critical military needs. ACTDs themselves are not acquisition programs, although they are designed to provide a residual, usable capability upon completion. If the user determines that additional units are needed beyond the residual capability and that these units can be funded, the additional buys shall constitute an acquisition program with an acquisition category generally commensurate with the dollar value and risk of the additional buy.

Automated Information System (AIS)

A combination of computer hardware and software, data, or telecommunications, that performs functions such as collecting, processing, transmitting, and displaying information. Excluded are computer resources, both hardware and software, that are physically part of, dedicated to, or essential in real time to the mission performance of weapon systems.

Commercial and Non-Developmental Items

Market research and analysis shall be conducted to determine the availability and suitability of existing commercial and non-developmental items prior to the commencement of a development effort, during the development effort, and prior to the preparation of any product description. For ACAT I and IA programs, while few commercial items meet requirements at a system level, numerous commercial components, processes, and practices have application to DoD systems.

Demilitarization and Disposal

At the end of its useful life, a system must be demilitarized and disposed of. During demilitarization and disposal, the PM shall ensure materiel determined to require demilitarization is controlled and shall ensure disposal is carried out in a way that minimizes DoD's liability due to environmental, safety, security, and health issues.

Developmental Test and Evaluation (DT&E)

DT&E shall identify potential operational and technological capabilities and limitations of the alternative concepts and design options being pursued; support the identification and description of design technical risks; and provide data and analysis in support of the decision to certify the system ready for operational test and evaluation.

Joint Program Management

Any acquisition system, subsystem, component or technology program that involves a strategy that includes funding by more than one DoD component during any phase of a system's life cycle shall be defined as a joint program. Joint programs shall be consolidated and collocated at the location of the lead Component's program office, to the maximum extent practicable.

Live Fire Test and Evaluation (LFT&E)

LFT&E must be conducted on a covered system, major munition program, missile program, or product improvement to a covered system, major munition program, or missile program before it can proceed beyond low-rate initial production. A covered system is any vehicle, weapon platform, or conventional weapon system that includes features designed to provide some degree of protection to users in combat and that is an ACAT I or II program. Depending upon its intended use, a commercial or non-developmental item may be a covered system, or a part of a covered system. (Change 4, 5000.2-R) Systems requiring LFT&E may not proceed beyond low-rate initial production until realistic survivability or lethality testing is completed and the report required by statute is submitted to the prescribed congressional committees.

Low Rate Initial Production (LRIP)

The objective of this activity is to produce the minimum quantity necessary to provide production- configured or representative articles for operational tests, establish an initial production base for the system; and permit an orderly increase in the production rate for the system, sufficient to lead to full-rate production upon successful completion of operational testing.

Major Automated Information System (MAIS) Acquisition Program

An AIS acquisition program that is (1) designated by ASD (C3I) as a MAIS, or (2) estimated to require program costs in any single year in excess of $30 million in FY96 constant dollars, total program costs in excess of $120 million in FY96 constant dollars, or total life-cycle costs in excess of $360 million in FY96 constant dollars. MAISs do not include highly sensitive classified programs.

Major Defense Acquisition Program (MDAP)

An acquisition program that is not a highly sensitive classified program (as determined by the Secretary of Defense) and that is: (1) designated by the Under Secretary of Defense (Acquisition and Technology) (USD [A&T]) as an MDAP, or (2) estimated by the USD (A&T) to require an eventual total expenditure for research, development, test and evaluation of more than $355 million in FY96 constant dollars or, for procurement, of more than $2.135 billion in FY96 constant dollars.

Major Milestone

A major milestone is the decision point that separates the phases of an acquisition program. MDAP milestones include, for example, the decisions to authorize entry into the engineering and manufacturing development phase or full rate production. MAIS milestones may include, for example, the decision to begin program definition and risk reduction.

Major System
A combination of elements that shall function together to produce the capabilities required to fulfill a mission need, including hardware, equipment, software, or any combination thereof, but excluding construction or other improvements to real property. A system shall be considered a major system if it is estimated by the DoD Component Head to require an eventual total expenditure for RDT&E of more than $135 million in FY96 constant dollars, or for procurement of more than $640 million in FY96 constant dollars, or if designated as major by the DoD Component Head.

Milestone Decision Authority (MDA)
The individual designated in accordance with criteria established by the USD (A&T), or by the ASD (C3I) for AIS acquisition programs, to approve entry of an acquisition program into the next phase.

Modifications
Any modification that is of sufficient cost and complexity that it could itself qualify as an ACAT I or ACAT IA program shall be considered for management purposes as a separate acquisition effort. Modifications that do not cross the ACAT I or IA threshold shall be considered part of the program being modified, unless the program is no longer in production. In that case, the modification shall be considered a separate acquisition effort. (Added from 5000.2-R)

Operational Support
The objectives of this activity are the execution of a support program that meets the threshold values of all support performance requirements and sustainment of them in the most life-cycle cost-effective manner. A follow-on operational testing program that assesses performance and quality, compatibility, and interoperability, and identifies deficiencies shall be conducted, as appropriate. This activity shall also include the execution of operational support plans, to include the transition from contractor to organic support, if appropriate. (Added from 5000.2-R)

Operational Test and Evaluation (OT&E)
OT&E shall be structured to determine the operational effectiveness and suitability of a system under realistic conditions (e.g., combat) and to determine if the operational performance requirements have been satisfied. The following procedures are mandatory: threat or threat representative forces, targets, and threat countermeasures, validated in coordination with DIA, shall be used; typical users shall operate and maintain the system or item under conditions simulating combat stress and peacetime conditions; the independent operational test activities shall use production or production representative articles for the dedicated phase of OT&E that supports the full-rate production decision, or for ACAT IA or other acquisition programs, the deployment decision; and the use of modeling and simulation shall be considered during test planning. There are more mandatory procedures (9 total) in 5000.2-R.

For additional information on acquisition terms, or terms not defined, please refer to AR 70-1, Army Acquisition Policy, available on the Web at http://www.army.mil/usapa/epubs/pdf/r70_1.pdf; or DA PAM 70-3, Army Acquisition Procedures, available on the Web at http://www.army.mil/usapa/epubs/pdf/p70_3.pdf.

Systems by Contractors

AAI Corp.
Extended Range Multi-Purpose (ERMP) Warrior Unmanned Aircraft System (UAS)
Tactical Unmanned Aerial Vehicle (TUAV)

Action Manufacturing
2.75" Family of Rockets

AcuSoft Inc.
One Semi-Automated Forces (OneSAF) Objective System

The Aegis Technology Group, Inc.
Joint Land Component Constructive Training Capability (JLCCTC)
One Semi-Automated Forces (OneSAF) Objective System

Aerial Machine and Tool, Inc.
Air Warrior

Aerojet
Army Tactical Missile System (ATACMS)
Guided Multiple Launch Rocket System (GMLRS)
Joint Common Missile (JCM)
Non-Line of Sight-Launch System (NLOS-LS)

Aerojet General
Tube-Launched, Optically-Tracked, Wire-Guided (TOW) Missiles

Airflyte Electronics Co.
Armored Knight

Akron Brass
Tactical Fire Fighting Truck (TFFT)

Alliant Techsystems
2.75" Family of Rockets
Artillery Ammunition
Medium Caliber Ammunition
Precision Guided Mortar Munitions (PGMM)
Small Caliber Ammunition
Tank Ammunition

Allison
Family of Medium Heavy Expanded Mobility Tactical Truck (HEMTT) and HEMTT Extended Service Program (ESP)
Family of Medium Tactical Vehicles (FMTV)

American Eurocopter
Light Utility Helicopter

American Ordnance
Artillery Ammunition
Medium Caliber Ammunition

AM General
High Mobility Multipurpose Wheeled Vehicle (HMMWV)
Improved Ribbon Bridge (IRB)

AMTEC Corp.
Medium Caliber Ammunition

Anniston Army Depot
Paladin/FAASV

AOT
Tank Ammunition

APC
Air/Missile Defense Planning and Control System (AMDPCS)

APPTIS
Combat Service Support Communications

Armacel Armor
Interceptor Body Armor (IBA)

Armor Holdings, Inc.
Air Warrior
High Mobility Multipurpose Wheeled Vehicle (HMMWV)
Family of Medium Tactical Vehicles (FMTV)

Armor Holdings, Aerospace & Defense Group
High Mobility Artillery Rocket System (HIMARS)

Armor Holdings/Simula Safety Systems, Inc.
Interceptor Body Armor (IBA)

Armor Holdings/Specialty Defense Systems
Interceptor Body Armor (IBA)

Armor Holdings TVS
Family of Medium Tactical Vehicles (FMTV)

ArmorWorks
Interceptor Body Armor (IBA)

Armtec Defense
Artillery Ammunition

Arvin/Meritor
Family of Medium Tactical Vehicles (FMTV)

AT&T Government Solutions
One Tactical Engagement Simulation System (OneTESS)

ATI Firth Sterling
Black Hawk/UH-60

ATK
Tube-Launched, Optically-Tracked, Wire-Guided (TOW) Missiles

Aviation Applied Technologies Directorate
Army Airborne Command and Control System (A2C2S)

Avon Protection Systems
Joint Service General Purpose Mask (JSGPM)

BAE Systems
Advanced Threat Infrared Counter measures (ATIRCM)/Common Missile Warning System (CMWS)
Airborne Reconnaissance Low (ARL)
Countermine

Army Data Distribution System (ADDS)/ Enhanced Position Location Reporting System (EPLRS)
Bradley Upgrade
Joint Tactical Ground Station (JTAGS)
Lightweight 155mm Howitzer (LW 155)
Paladin/FAASV
Precision Guided Mortar Munitions (PGMM)
Thermal Weapon Sight II
Warfighter Information Network-Tactical (WIN-T)

BAE Systems Bofors Defense
Excalibur (XM982)

BAE Systems Electronics & Integrated Solutions
2.75" Family of Rockets

Barrett Firearms Manufacturing
Sniper Systems

Bechtel Aberdeen
Chemical Demilitarization

Bell Helicopter
Kiowa Warrior

Bell Helicopter Textron
Armed Reconnaissance Helicopter (ARH)

Boeing
Chinook/CH-47F Improved Cargo Helicopter (ICH)
Longbow Apache
PATRIOT (PAC-3)
Surface Launched Advanced Medium Range Air-to-Air Missile (SLAMRAAM)

Boeing Satellite Systems
Defense Enterprise Wideband SATCOM System (DEWSS)

Buffalo Turbine, LLC
Rapid Equipping Force (REF)

Bruhn NewTech
Joint Warning and Reporting Network (JWARN)

CACI Technologies
Airborne Reconnaissance Low (ARL)
Guardrail Common Sensor (GR/CS)
Nuclear Biological Chemical Reconnaissance Vehicle (NBCRV)-Stryker

CAE USA
Light Utility Helicopter (LUH)

Capewell
Joint Precision Airdrop System (JPADS)

Carleton Technologies, Inc.
Air Warrior (AW)

CAS, Inc.
Forward Area Air Defense Command and Control (FAAD C2)
Joint Land Attack Cruise Missile Defense (LACMD) Elevated Netted Sensor System (JLENS)

Caterpillar
Family of Medium Tactical Vehicles (FMTV)

CECOM Software Engineering Center
Maneuver Control System (MCS)

Ceradyne, Inc.
Interceptor Body Armor (IBA)

CDW-G
Medical Communications for Combat Casualty Care (MC4)

Chenega Integrated Systems, LLC
Unit Water Pod System (Camel)

Cisco
Common Hardware Systems (CHS)

Colt's Manufacturing
Small Arms

Computer Giant
Combat Service Support Communications

Computer Sciences Corp.
Advanced Field Artillery Tactical Data System (AFATDS)
Close Combat Tactical Trainer (CCTT)
Land Warrior
Physical Security Force
Protection System of Systems
Transportation Coordinators' Automated Information for Movement System II (TC-AIMS II)

CMC Electronics, Cincinnati
Lightweight Laser Designator Rangefinder (LLDR)

CMI
Tactical Unmanned Aerial Vehicle (TUAV)

COMTECH Mobile Datacom
Movement Tracking System (MTS)
Force XXI Battle Command Brigade-and-Below (FBCB2)

Condor Pacific
Tube-Launched, Optically-Tracked, Wire-Guided (TOW) Missiles

Crossroads Industrial Services
Joint Combat Identification Marking System (JCIMS)

CSC
Integrated System Control (ISYSCON) (V)4/Tactical Internet Management System (TIMS)

Cubic Simulation Systems (formerly ECC International Corp.)
Engagement Skills Trainer (EST) 2000

Cummins Power
Forward Repair System (FRS)

CyTerra Corp.
Countermine

Defiance
High Mobility Multipurpose Wheeled Vehicle (HMMWV)

Detroit Diesel
Heavy Expanded Mobility Tactical Truck (HEMTT) and HEMTT Extended Service Program (ESP)
Line Haul Tractor

Deutz U.S.A
Tactical Fire Fighting Truck (TFFT)

Dewey Electronics
Tactical Electrical Power (TEP)

DPA
Tactical Unmanned Aerial Vehicle (TUAV)

DRS-ESSI
Tactical Electrical Power (TEP)

DRS Fermont
Tactical Electrical Power (TEP)

DRS Technologies
Chemical Biological Protective Shelter (CBPS)
Common Hardware Systems (CHS)
Force XXI Battle Command Brigade-and-Below (FBCB2)
Integrated Family of Test Equipment (IFTE)
Kiowa Warrior
Thermal Weapon Sight II

DRS Laurel Technologies
Armored Knight

DRS Radian, Inc.
Physical Security Force Protection System of Systems

DRS Sustainment Systems, Inc.
Armored Knight
Modular Fuel System (MFS)

DRS Tactical Systems
Armored Knight

DSE Corp.
Medium Caliber Ammunition

Ducommun AeroStructures
Longbow Apache

DRS Technologies

DynCorp International
Advanced Threat Infrared Counter measures (ATIRCM)/Common Missile Warning System (CMWS)
Fixed Wing

EADS North America
Light Utility Helicopter (LUH)

Eagle Industries, Inc.
Joint Combat Identification Marking System (JCIMS)

E.D. Etnyre and Co.
Modular Fuel System (MFS)

EDO Corp.
XM307

EFW
Bradley Upgrade

EG&G Technical Services, Inc.
Chemical Demilitarization
Physical Security Force Protection System of Systems

Engineering Professional Services
Advanced Field Artillery Tactical Data System (AFATDS)
Army Data Distribution System (ADDS)/ Enhanced Position Location Reporting System (EPLRS)

Engineering Solutions & Products Inc.
Force XXI Battle Command Brigade-and-Below (FBCB2)
Global Command and Control System-Army (GCCS-A)
Integrated System Control (ISYSCON) (V)4/Tactical Internet Management System (TIMS)

European Aeronautic Defence and Space Company (EADS)
Medium Extended Air Defense System (MEADS)

EyakTek
Combat Service Support Communications

Fabrique National Manufacturing, LLC
Small Arms

Fairfield
Distributed Learning System (DLS)

FBM Babcock Marine
Improved Ribbon Bridge (IRB)

FC Business Systems
Global Command and Control System–Army (GCCS-A)

FLIR Systems, Inc.
Armed Reconnaissance Helicopter (ARH)

Force XXI Battle Command Brigade and Below
Movement Tracking System (MTS)

Freightliner Trucks
Line Haul Tractor

GEP
High Mobility Multipurpose Wheeled Vehicle (HMMWV)

GenCorp
Joint Tactical Ground Stations (JTAGS)

General Atomics
Extended Range Multi-Purpose (ERMP) Warrior Unmanned Aircraft System (UAS)
Rapid Equipping Force (REF)

General Dynamics
Abrams Upgrade
Advanced Field Artillery Tactical Data System (AFATDS)
All Source Analysis System (ASAS)
Common Hardware Systems (CHS)
Forward Area Air Defense Command and Control (FAAD C2)
Global Command and Control System-Army (GCCS-A)
Joint Combat Missile (JCM)
Lightweight 155mm Howitzer (LW 155)

Maneuver Control System (MCS)
Medical Communications for Combat Casualty Care (MC4)
Mounted Warrior
Stryker
Tactical Unmanned Aerial Vehicle (TUAV)
Warfighter Information Network–Tactical (WIN-T)

General Dynamics ATP Division
Joint Biological Point Detection System (JBPDS)
Small Arms

General Dynamics C4 Systems, Inc.
Air Warrior
Land Warrior (LW)
Prophet

General Dynamics Armament and Technical Products
2.75" Family of Rockets
XM307

General Dynamics Land Systems
Land Warrior (LW)
Nuclear Biological Chemical Reconnaissance Vehicle (NBCRV)–Stryker

General Dynamics Ordnance and Tactical Systems
2.75" Family of Rockets
Artillery Ammunition
Conventional Ammunition Demilitarization
Excalibur (XM982)
Medium Caliber Ammunition
Mortar Systems
Small Caliber Ammunition
Tank Ammunition

General Dynamics Robotics Systems
Physical Security Force Protection System of Systems

General Dynamics Santa Barbara
Improved Ribbon Bridge (IRB)

General Electric
Black Hawk

GESTALT
Global Command and Control System-Army (GCCS-A)

Global Communications Solutions
Combat Service Support Communications

Global Secure Corp.
Rapid Equipping Force (REF)

GM
High Mobility Multipurpose Wheeled Vehicle (HMMWV)

Goodrich-Hella
Army Tactical Missile System (ATACMS)
Black Hawk

Grove Worldwide
Forward Repair System (FRS)

GTSI
Combat Service Support Communications
Global Command and Control System-Army (GCCS-A)
Integrated System Control (ISYSCON) (V)4/Tactical Internet Management System (TIMS)
Medical Communications for Combat Casualty Care (MC4)

Gulfstream
Fixed Wing

Gyrocam Systems
Rapid Equipping Force (REF)

Gyrocam Systems LLC
Countermine

Harris Corp.
Defense Enterprise Wideband SATCOM System (DEWSS)
High Mobility Artillery Rocket System (HIMARS)
Warfighter Information Network-Tactical (WIN-T)

Heckler and Koch Defense, Inc.
Small Arms

HELLFIRE LLC
Hellfire Family of Missiles

Hewlett-Packard
Common Hardware Systems (CHS)

Holland Hitch
Line Haul Tractor

Honeywell Aerospace Electronics
Abrams Upgrade
Army Tactical Missile System (ATACMS)
Chinook/CH-47F Improved Cargo Helicopter (ICH)
Guided Multiple Launch Rocket System (GMLRS)
Mortar Systems

Honeywell ES&S
Armed Reconnaissance Helicopter (ARH)

Honeywell, Inc.
Kiowa Warrior

Howmet Castings
Lightweight 155mm Howitzer (LW 155)

Idaho Technologies
Joint Biological Agent Identification Diagnostic System (JBAIDS)

Indigo Systems
Lightweight Laser Designator Rangefinder (LLDR)

Information Systems Support, Inc.
Army Key Management System (AKMS)

Ingersoll-Rand
Forward Repair System (FRS)

Innolog
Army Data Distribution System (ADDS)/ Enhanced Position Location Reporting System (EPLRS)

Inter4
Army Key Management System (AKMS)

Intercoastal Electronics
Improved Target Acquisition System (ITAS)

International Business Machines (IBM)
Distributed Learning System (DLS)

Intertek Laboratories
XM307

ITT Industries
Defense Enterprise Wideband SATCOM System (DEWSS)
Night Vision Devices

JANUS Research
Secure Mobile Anti-Jam Reliable Tactical-Terminal (SMART-T)

Johns Hopkins University Applied Physics Laboratory
Defense Enterprise Wideband SATCOM System (DEWSS)
Medical Communications for Combat Casualty Care (MC4)

Joint Venture Yulista SES-I
Army Airborne Command and Control System (A2C2S)

Kaegan Corp.
Close Combat Tactical Trainer (CCTT)

Kaman
Army Tactical Missile System (ATACMS)

Kidde Dual Spectrum
Paladin/FAASV

King Aerospace
Fixed Wing

Knight's Armaments Co.
Sniper Systems

L-3 Communications
Tank Ammunition

L-3 Communications Titan Group
Advanced Threat Infrared Counter measures (ATIRCM)/Common Missile Warning System (CMWS)
Aviation Combined Arms Tactical Trainer (AVCATT)
Battle Command Sustainment Support System (BCS3)
Bradley Upgrade
Extended Range Multi-Purpose (ERMP) Warrior Unmanned Aircraft System (UAS)
Guardrail Common Sensor (GR/CS)
Global Combat Support System-Army (GCSS-Army)
Medical Communications for Combat Casualty Care (MC4)
Non-Line of Sight-Launch System (NLOS-LS)
Physical Security Force Protection System of Systems
Prophet
Tank Ammunition
Transportation Coordinators' Automated Information for Movement System II (TC-AIMS II)

L-3 Communications Interstate Electronics Corp.
Excalibur (XM982)

L-3 Communications Space and Navigation
High Mobility Artillery Rocket System (HIMARS)

L3/IAC
Non-Line of Sight-Launch System (NLOS-LS)

L3 Titan Corp.
Distributed Learning System (DLS)

Laser Systems Division
Lightweight Laser Designator Rangefinder (LLDR)

Letterkenny Army Depot
High Mobility Multipurpose Wheeled Vehicle (HMMWV)

Lenkflugkorpersysteme (LFK)
Medium Extended Air Defense System (MEADS)

Lincoln Labs
Secure Mobile Anti-Jam Reliable Tactical-Terminal (SMART-T)

Litton Advanced Systems
Airborne Reconnaissance Low (ARL)

Lockheed Martin
Airborne Reconnaissance Low (ARL)
Army Tactical Missile System (ATACMS)
Battle Command Sustainment Support System (BCS3)
Close Combat Tactical Trainer (CCTT)
Guardrail Common Sensor (GR/CS)
Guided Multiple Launch Rocket System (GMLRS)
Global Command and Control System-Army (GCCS-A)
High Mobility Artillery Rocket System (HIMARS)
Javelin
Joint Common Missile (JCM)
Joint Tactical Ground Station (JTAGS)
Longbow Apache
Maneuver Control System (MCS)
Medium Extended Air Defense System (MEADS)
One Semi-Automated Forces (OneSAF) Objective System
PATRIOT (PAC-3)

Lockheed Martin Baltimore
Non-Line of Sight-Launch System (NLOS-LS)

Lockheed Martin Dallas
Non-Line of Sight-Launch System (NLOS-LS)

Lockheed Martin Information Systems
Joint Land Component Constructive Training Capability (JLCCTC)

Lockheed Martin Integrated Systems, Inc.
All Source Analysis System (ASAS)

Lockheed Martin Mission Systems
Warfighter Information Network-Tactical (WIN-T)

Longbow LLC
Hellfire Family of Missiles

LTI DATACOM
Combat Service Support Communications

M7 Aerospace
Fixed Wing

Main Military Authority
High Mobility Multipurpose Wheeled Vehicle (HMMWV)

MANTECH
All Source Analysis System (ASAS)

Marvin Land Systems
Paladin/FAASV

MBDA-Italia
Medium Extended Air Defense System (MEADS)

MCII
Tactical Electrical Power (TEP)

Mechanical Equipment Co.
Lightweight Water Purification System (LWP)

Meritor
Line Haul Tractor

Mevatec
Joint Tactical Ground Station (JTAGS)

MICOR Industries, Inc.
XM101 Common Remotely Operated Weapon Station (CROWS)

Michelin
Heavy Expanded Mobility Tactical Truck (HEMTT) and HEMTT Extended Service Program (ESP)

Mil-Mar Century, Inc.
Load Handling System Compatible Water Tank Rack (Hippo)

MMIST
Joint Precision Airdrop System (JPADS)

Moog
Tube-Launched, Optically-Tracked, Wire-Guided (TOW) Missiles

Night Vision Equipment Company
Joint Combat Identification Marking System (JCIMS)

Nordic Ammunition
Small Caliber Ammunition

Northrop Grumman
Air/Missile Defense Planning and Control System (AMDPCS)
Battle Command Sustainment Support System (BCS3)
Countermine
Defense Enterprise Wideband SATCOM System (DEWSS)
Distributed Common Ground System-Army (DCGS-A)
Guardrail Common Sensor (GR/CS)
Integrated Family of Test Equipment (IFTE)
Joint Tactical Ground Stations (JTAGS)
Longbow Apache
Maneuver Control System (MCS)
Movement Tracking System (MTS)
Night Vision Devices
Paladin/FAASV
Tactical Operations Center (TOC)
Transportation Coordinators' Automated Information for Movement System II (TC-AIMS II)

Northrop Grumman Electronic Systems Laser Systems Division
Lightweight Laser Designator Rangefinder (LLDR)

Northrop Grumman Information Technology
Joint Warning and Reporting Network (JWARN)
One Semi-Automated Forces (OneSAF) Objective System

Northrop Grumman Mission Systems
Forward Area Air Defense Command and Control (FAAD C2)
Global Combat Support System-Army (GCSS-Army)
Integrated System Control (ISYSCON) (V)4/Tactical Internet Management System (TIMS)

Northrop Grumman Space & Mission Systems Corp.
Force XXI Battle Command Brigade-and-Below (FBCB2)

Oshkosh Truck Corp.
Dry Support Bridge (DSB)
Heavy Expanded Mobility Tactical Truck (HEMTT) and HEMTT Extended Service Program (ESP)
Improved Ribbon Bridge (IRB)
Tactical Fire Fighting Truck (TFFT)

Omega Training Group
Land Warrior

Oppenheimer
Armored Knight

Pacific Scientific
Precision Guided Mortar Munitions (PGMM)

Parsons Infrastructure & Technology
Chemical Demilitarization

Penn State University
Meteorological Measuring Set–Profiler (MMS-P)

Phoenix Coaters LLC
Tactical Electrical Power (TEP)

Point Blank Body Armor
Interceptor Body Armor (IBA)

Precise Industries
Family of Medium Tactical Vehicles (FMTV)

Precision Castparts Corp.
Lightweight 155mm Howitzer (LW 155)

Protective Materials Company
Interceptor Body Armor (IBA)

Prototype Integration Facility
Army Airborne Command and Control System (A2C2S)

Rauch, Inc.
Joint Combat Identification Marking System (JCIMS)

Raytheon
Abrams Upgrade
Advanced Field Artillery Tactical Data System (AFATDS)
Army Data Distribution System (ADDS)/

Enhanced Position Location Reporting System (EPLRS)
Bradley Upgrade
Excalibur (XM982)
Guardrail Common Sensor (GR/CS)
Improved Target Acquisition System (ITAS)
Javelin
Joint Land Attack Cruise Missile Defense (LACMD) Elevated Netted Sensor System (JLENS)
Land Warrior
Non-Line of Sight-Launch System (NLOS-LS)
PATRIOT (PAC-3)
Secure Mobile Anti-Jam Reliable Tactical-Terminal (SMART-T)
Surface Launched Advanced Medium Range Air-to-Air Missile (SLAMRAAM)
Tank Ammunition
Tube-Launched, Optically-Tracked, Wire-Guided (TOW) Missiles
XM307

Raytheon Technical Services, Inc.
Air Warrior (AW)

Recon Optical, Inc.
XM101 Common Remotely Operated Weapon Station (CROWS)

Red River Army Depot
High Mobility Multipurpose Wheeled Vehicle (HMMWV)

REMEC
Joint Common Missile (JCM)

Robertson Aviation
Chinook/CH-47F Improved Cargo Helicopter (ICH)

Rock Island Arsenal
Forward Repair System (FRS)

Rockwell Collins
Armed Reconnaissance Helicopter (ARH)
Black Hawk
Chinook/CH-47F Improved Cargo Helicopter (ICH)
Close Combat Tactical Trainer (CCTT)
Global Positioning System (GPS)
Prophet

Rolls Royce Corp.
Kiowa Warrior

SAIC
Army Key Management System (AKMS)
Installation Protection Program (IPP) Family of Systems
Joint Network Management Systems (JNMS)
One Semi-Automated Forces (OneSAF) Objective System

Science & Engineering Services, Inc.
Integrated Family of Test Equipment (IFTE)

Sechan Electronics
Secure Mobile Anti-Jam Reliable Tactical-Terminal (SMART-T)

Secure Communications Systems, Inc.
Air Warrior (AW)

Segovia
Combat Service Support Communications

Sierra Nevada Corp.
Airborne Reconnaissance Low (ARL)
Army Key Management System (AKMS)
Tactical Unmanned Aerial Vehicle (TUAV)

Signal Solutions
Combat Service Support Communications

Sikorsky Aircraft Corp.
Advanced Threat Infrared Countermeasures (ATIRCM)/Common Missile Warning System (CMWS)
Light Utility Helicopter (LUH)

Simula Safety Systems, Inc.
Air Warrior (AW)
Interceptor Body Armor (IBA)

Sistemas
Improved Ribbon Bridge (IRB)

Skillsoft
Distributed Learning System (DLS)

Smiths
Kiowa Warrior

Smiths Detection
Chemical Biological Protective Shelter (CBPS)
Joint Chemical Agent Detector (JCAD)
Meteorological Measuring Set-Profiler (MMS-P)

Snap-on Industrial
Forward Repair System (FRS)

SNC Technologies
Artillery Ammunition

Specialty Defense Systems
Interceptor Body Armor (IBA)

Spincraft
Army Tactical Missile System (ATACMS)

Strong Enterprises
Joint Precision Airdrop System (JPADS)

Sun Microsystems
Common Hardware Systems (CHS)

Sypris
Army Key Management System (AKMS)

Tactical Support Equipment
Rapid Equipping Force (REF)

TallaTech
Common Hardware Systems (CHS)
Mortar Systems

TAMSCO
Combat Service Support Communications

Tapestry Solutions
Battle Command Sustainment Support System (BCS3)
Joint Land Component Constructive Training Capability (JLCCTC)

TCOM
Joint Land Attack Cruise Missile

Defense (LACMD) Elevated Netted Sensor System (JLENS)

Technical Products
XM307

Tecom
Tactical Unmanned Aerial Vehicle (TUAV)

Telephonics Corp.
Air Warrior (AW)

Textron Marine & Land Systems
Armored Knight
Armored Security Vehicle (ASV)

Thales
Tube-Launched, Optically-Tracked, Wire-Guided (TOW) Missiles

Thales Raytheon Systems
Sentinel

Titan Corp.
Advanced Field Artillery Tactical Data System (AFATDS)

Triumph Systems Los Angeles
Lightweight 155mm Howitzer (LW 155)

United Plastics Fabricators
Tactical Fire Fighting Truck (TFFT)

United Technologies
Black Hawk

U.S. Army Information Systems Engineering Command
Defense Enterprise Wideband SATCOM System (DEWSS)

Vertu Corp.
Small Arms

Vickers
High Mobility Artillery Rocket System (HIMARS)

Viecore
Maneuver Control System (MCS)

Vision Technology Miltope Corporation
Integrated Family of Test Equipment (IFTE)

XMCO
Dry Support Bridge (DSB)

Washington Demilitarization Company
Chemical Demilitarization

Washington Group International
Chemical Demilitarization

WESCAM
Airborne Reconnaissance Low (ARL)

Westwind Technologies, Inc.
Advanced Threat Infrared Counter measures (ATIRCM)/Common Missile Warning System (CMWS)
Air Warrior (AW)

WEW Westerwalder Eisenwerk
Load Handling System Compatible Water Tank Rack (Hippo)

Wexford Group International
Battle Command Sustainment Support System (BCS3)

Williams Fairey Engineering, Ltd.
Dry Support Bridge (DSB)

Wolf Coach, Inc., an L-3 Company
Analytical Laboratory System-System Enhancement Program (ALS-SEP)
Unified Command Suite (UCS)

W.S. Darley Corp.
Tactical Fire Fighting Truck (TFFT)

ZETA
Guardrail Common Sensor (GR/CS)

Contractors by State

Alabama
Anniston Army Depot
ATI Firth Sterling
Boeing
CAS, Inc.
General Dynamics Land Systems
General Dynamics Ordnance and Tactical Systems
Washington Group International
Westwind Technologies, Inc.

Arizona
ArmorWorks
Boeing
General Dynamics C4 Systems, Inc.
Honeywell, Inc.
Lockheed Martin
Raytheon
Robertson Aviation

Arkansas
Washington Demilitarization Company

California
Aerial Machine and Tool, Inc.
Aerojet
Armacel Armor
Armtec Defense
BAE
Computer Sciences Corp.
DRS
DuoCommon Aerostructures
General Atomics
General Dynamics Ordnance and Tactical Systems
Indigo Systems
Kidde Dual Spectrum
L-3 Communications
L-3 Communications Interstate Electronics Corp.
L3 Titan Corp.
L3/IAC
Lockheed Martin
Marvin Land Systems
Northrop Grumman
Northrop Grumman Information Technology
Northrop Grumman Mission Systems
Northrop Grumman Space & Mission Systems Corp.
Pacific Scientific
Raytheon
REMEC
Tapestry Solutions
Tecom
Thales Raytheon Systems
Triumph Systems Los Angeles

Colorado
GenCorp

Connecticut
Capewell
Colt's Manufacturing
Condor Pacific
Kaman
United Technologies

Florida
AcuSoft, Inc.
Alliant Techsystems
AT&T Government Solutions
CAE USA
Chenega Integrated Systems, LLC
Cubic Simulation Systems
DRS
DRS Tactical Systems
DRS Technologies
DSE (Balimoy) Corp.
General Dynamics
General Dynamics Ordnance and Tactical Systems
Goodrich-Hella
Gyrocam Systems
Harris Corp.
Honeywell
Kaegan Corp.
Knight's Armaments Co.
L-3 Communications
Lockheed Martin
Lockheed Martin Information Systems
MEADS International
Northrop Grumman Electronic Systems Laser Systems Division
Northrop Grumman Information Technology
Point Blank Body Armor
Protective Materials Company
Raytheon
SAIC
Strong Enterprises
Talla-Tech
Thales Raytheon Systems
The Aegis Technology Group, Inc.

Georgia
Deutz U.S.A.
FC Business Systems
Gulfstream
Omega Training Group

Illinois
CDW-G
E.D. Etnyre and Co.
General Dynamics Ordnance and Tactical Systems
Rock Island Arsenal
Snap-on Industrial
W.S. Darley Corp.

Indiana
Parsons Infrastructure and Technology
Rauch, Inc.
Raytheon
Raytheon Technical Services, Inc.

Iowa
American Ordnance
Rockwell Collins

Kentucky
Ingersoll-Rand
L-3 Communications

Louisiana
Mechanical Equipment Co.
Textron Marine & Land Systems

Maine
General Dynamics ATP

Maryland
AAI Corp.
Bechtel Aberdeen
Bruhn NewTech
COMTECH
COMTECH Mobile Datacom

Johns Hopkins University Applied Physics Laboratory
Lockheed Martin Baltimore
Lockheed Martin Mission Systems
Naval Air Warfare Center Aircraft Division
Northrop Grumman
Rockwell Collins
SFA Frederick Manufacturing
Smiths Detection, Inc.

Massachusetts
BAE
General Dynamics
General Dynamics C4S
General Electric
Raytheon
Wolf Coach, Inc.

Michigan
Avon Protection Systems
General Dynamics
General Dynamics Land Systems
L-3 Communications
Meritor
Oshkosh Truck Corp.
Precision Castparts Corp.

Minnesota
Alliant Techsystems
AM General
Cummins Power

Mississippi
American Eurocopter
BAE Systems
Raytheon
Thales Raytheon Systems
Vickers

Missouri
Alliant Techsystems
DRS Sustainment Systems, Inc.

Nevada
Sierra Nevada Corp.

New Hampshire
BAE Systems
BAE Systems Electronics & Integrated Solutions
Skillsoft

New Jersey
BAE Systems
CACI
Computer Sciences Corp.
DRS Technologies
Engineering Professional Services
Engineering Solutions & Products, Inc.
Force XXI Battle Command Brigade and Below (FBCB2)
GESTALT
Innolog
Intertech Laboratories
JANUS Research
L-3 Communications Space and Navigation
Lockheed Martin
Lockheed Martin Integrated Systems
SAIC
Viecore

New Mexico
Aerojet
Honeywell Aerospace Electronics

New York
Buffalo Turbine, LLC
EDO Corp.
Lockheed Martin
Telephonics Corp.

North Carolina
General Dynamics ATP Division
Tactical Support Equipment

Ohio
Akron Brass
Arvin/Meritor
CMC Electronics, Cincinnati
General Dynamics
Mil-Mar Century, Inc.

Oklahoma
Titan Corp.

Oregon
FLIR Systems, Inc.
Freightliner Trucks
Precision Castparts Corp.
Washington Demilitarization Company

Pennsylvania
Action Manufacturing
BAE
Boeing
General Dynamics
General Dynamics Ordnance and Tactical Systems
Grove Worldwide
L-3 Communications
Night Vision Equipment Company
Penn State University
Sechan Electronics

South Carolina
Fabrique National Manufacturing, LLC

Tennessee
American Ordnance
AOT
Barrett Firearms Manufacturing

Texas
Armor Holdings TVS
Armor Holdings, Aerospace & Defense Group
American Eurocopter
Bell Helicopter Textron
DRS
DynCorp International
King Aerospace
L-3 Communications
Lockheed Dallas
Lockheed Martin
M7 Aerospace
MANTECH
Raytheon

Utah
EG&G Technical Services, Inc.
Idaho Technologies
L-3 Communications
Moog

Vermont
General Dynamics ATP Division
Goodrich

Virginia
Aerial Machine and Tool, Inc.
Aerojet
Alliant Techsystems
AT&T Government Solutions
ATK
CACI
Computer Sciences Corp.
CSC

DPA
EADS North America
Fairfield
FC Business Systems
General Dynamics
GTSI Corp.
Heckler and Koch Defense, Inc.
L-3 Communications
Lockheed Martin
Northrop Grumman Mission Systems
Raytheon
SAIC
Vertu Corp.
Wexford Group International
ZETA

West Virginia
Alliant Techsystems

Wisconsin
Alliant Techsystems
Oshkosh Truck Corp.
Spincraft

Points of Contact

2.75" Family of Rockets
JAMS Project Office
ATTN: SFAE-MSL-JAMS
Redstone Arsenal, AL 35898-8000

Abrams Upgrade
ATTN: SFAE-GCS-CS-A
6501 E. 11 Mile Road
Warren, MI 48397-5000

Advanced Field Artillery Tactical Data System (AFATDS)
Product Director
Fire Support Command and Control
ATTN: SFAE-C3T-BC-FSC2
Building 2525
Fort Monmouth, NJ 07703-5404

Advanced Threat Infrared Countermeasures (ATIRCM)/Common Missile Warning System (CMWS)
PM IRCM
Building 5309
Sparkman Center
Redstone Arsenal, AL 35898-5000

Aerial Common Sensor (ACS)
PM Army Airborne Command and Control System (PM A2C2S)
ATTN: SFAE-C3T-FBC-ACS
650 Discovery Drive
Huntsville, AL 35806

Air Warrior (AW)
PM Air Warrior
ATTN: SFAE-SDR-AW
Redstone Arsenal, AL 35898

Air/Missile Defense Planning and Control System (AMDPCS)
C-RAM Program Office
ATTN: SFAE-C3T-CR-AMD
Redstone Arsenal, AL 35898-5000

Airborne Reconnaissance Low (ARL)
ATTN: SFAE-IEWS-ACS
Building 288
Fort Monmouth, NJ 07703

All Source Analysis System (ASAS)
PD Intelligence Fusion
10115 Duportail Road
Fort Belvoir, VA 22060-5812

Analytical Laboratory System–System Enhancement Program (ALS-SEP)
ATTN: SFAE-CBD-WMDCSS
5183 Blackhawk Road
APG, MD 21010-5424

Armored Knight
PM HBCT
PM-Fire Support Platforms
ATTN: SFAE-GCS-HBCT-F
6501 East 11 Mile Rd
Warren, MI 48397-5000

Armored Security Vehicle (ASV)
PM Medium Tactical Vehicles
SFAE-CSS-MTVL
6501 Eleven Mile Road, MS 245
Warren, MI 48397-5000

Army Airborne Command and Control System (A2C2S)
PM Army Airborne Command and Control System
(PM A2C2S)
ATTN: SFAE-C3T-FBC-ACS
650 Discovery Drive
Huntsville, AL 35806

Army Data Distribution System (ADDS)/ Enhanced Position Location Reporting System (EPLRS)
PM Tactical Radio Systems
Building 456
Fort Monmouth, NJ 07703

Army Key Management System (AKMS)
PdM, NETOPS-CF
ATTN: SFAE-C3T-TRC-NETOPS-CF
Fort Monmouth, NJ 07703

Army Tactical Missile System (ATACMS)
Precision Fires Rocket and Missile Systems Project Office
ATTN: SFAE-MSL-PF-AT
Building 5250
Redstone Arsenal, AL 35898

Artillery Ammunition
PM Combat Ammunition Systems
ATTN: SFAE-AMO-CAS
Picatinny Arsenal, NJ 07806

Aviation Combined Arms Tactical Trainer (AVCATT)
Product Manager, Air and Command Tactical Trainers
12350 Research Parkway
Orlando, FL 32826-3276

Battle Command Sustainment Support System (BCS3)
PM Battle Command Sustainment Support System (BCS3)
ATTN: SFAE-C3T-GC-BCS3
10109 Gridley Road
Fort Belvoir, VA 22060

Black Hawk/UH-60)
Utility Helicopters
Project Manager
ATTN: SFAE-AV-UH
Building 5308
Redstone Arsenal, AL 35898

Bradley Upgrade
6501 East Eleven Mile Road
ATTN: SFAE-GCS-CS
Warren, MI 48397-5000

Chemical Biological Protective Shelter (CBPS)
Commander
Naval Sea Systems Command
ATTN: SEA 05P5
1333 Isaac Hall SE
Washington Navy Yard
Washington, DC 20376-5150

Chemical Demilitarization
Chemical Materials Agency (CMA)
ATTN: AMSCM-D
5183 Blackhawk Road
APG-EA, MD 21010-5424

Chinook/CH-47F Improved Cargo Helicopter (ICH)
PM Cargo Helicopters
ATTN: SFAE-AV-CH-ICH
Building 5678
Redstone Arsenal
Huntsville, AL 35898

Close Combat Tactical Trainer (CCTT)
Assistant Project Manager, Close Combat Tactical Trainers
12350 Research Parkway
Orlando, FL 32826-3276

Combat Service Support Communications
Combat Service Support Automated Information Systems Interface (CAISI) Office
6700 Springfield Center Dr., Suite E
Springfield, VA 22150

Common Hardware Systems (CHS)
Product Director Common Hardware Systems (PD CHS)
ATTN: SFAE-C3T-TRC-CHS
Bldg 457
Fort Monmouth, NJ 07703

Conventional Ammunition Demilitarization
PM Demilitarization
ATTN: SFAE-AMO-JS-D
Picatinny Arsenal, NJ 07806

Countermine
PM Countermine
ATTN: SFAE-AMO-CCS
Fort Belvoir, VA 22060-5811

Defense Enterprise Wideband SATCOM System (DEWSS)
PM Defense Communications and Army Transmission Systems
Building 209
Fort Monmouth, NJ 07703-5509

Distributed Common Ground System-Army (DCGS-A)
ATTN: SFAE-IEWS-DCGS-A
Building 550
Saltzman Avenue
Fort Monmouth, NJ 07703-5301

Distributed Learning System (DLS)
PM DLS
ATTN: SFAE-PS-DL
11846 Rock Landing Drive, Ste B
Newport News, VA 23606-4206

Dry Support Bridge (DSB)
PM Assured Mobility Systems
(SFAE-CSS-FP-E) MS 401
6501 East Eleven Mile Road
Warren, MI 48397-5000

Engagement Skills Trainer (EST) 2000
PM Combined Arms Tactical Trainer
12350 Research Parkway
Orlando, FL 32826-3276

Excalibur (XM982)
PM Combat Ammo Systems
ATTN: SFAE-AMO-CAS-EX
Picatinny Arsenal, NJ 07806

Extended Range Multi-Purpose (ERMP) Warrior Unmanned Aircraft System (UAS)
PM Unmanned Aircraft Systems
ATTN: SFAE-AV-UAS
Redstone Arsenal, AL 35898

Family of Medium Tactical Vehicles (FMTV)
ATTN: SFAE-CSS
6501 East Eleven Mile Road
Warren, MI 48397-5000

Fixed Wing
ATTN: SFAE-AV-AS-FW
Building 5309
Redstone Arsenal, AL 35898-5000

Force XXI Battle Command Brigade-and-Below (FBCB2)
PM, FBCB2
Building 2525, Bay 1
Fort Monmouth, NJ 07703-5408

Forward Area Air Defense Command and Control (FAAD C2)
C-RAM Program Office
ATTN: SFAE-C3T-CR
Redstone Arsenal, AL 35898-5000

Forward Repair System (FRS)
PM SKOT–Forward Repair System (FRS)
ATTN: AMSTA-LC-CTT-M
Building 104, 1st Floor, East Wing
Rock Island, IL 61299-7630

Global Combat Support System–Army (GCSS-Army)
PM SALE
9350 Hall Road
Fort Belvoir, VA 22060-5526

Global Command and Control System–Army (GCCS-A)
PEO C3T
ATTN: SFAE-C3T-GC
Building 5100
Fort Monmouth, NJ 07703-5404

Global Positioning System (GPS)
ATTN: GPSW/GPA
483 North Aviation Blvd
El Segundo, CA 90245

Guardrail Common Sensor (GR/CS)
SFAE-IEWS-ACS
Building 288
Fort Monmouth, NJ 07703

Guided Multiple Launch Rocket System (GMLRS)
Precision Fires Rocket and Missile Systems Project Office
ATTN:SFAE-MSL-PF-PGM/R
Building 5250
Redstone Arsenal, AL 35898

Heavy Expanded Mobility Tactical Truck (HEMTT) and HEMTT Extended Service Program (ESP)
ATTN: SFAE-CSS-TV-H
MS 429
6501 East Eleven Mile Road
Warren, MI 48397-5000

Hellfire Family of Missiles
JAMS Project Office
ATTN: SFAE-MSLS-JAMS
Redstone Arsenal, AL
35898-8000

High Mobility Artillery Rocket System (HIMARS)
Precision Fires Rocket and Missile Systems Project Office
ATTN: SFAE- MSL- PF-FAL
Building 5250
Redstone Arsenal, AL 35898

High Mobility Multipurpose Wheeled Vehicle (HMMWV)
Product Manager Light Tactical Vehicles
ATTN: SFAE-CSS-TV-L
6501 Eleven Mile Road, MS 245
Warren, MI 48397-5000

Improved Ribbon Bridge (IRB)
Product Manager
Assured Mobility Systems (SAFE-CSS-FP-C) MS 401
6501 East Eleven Mile Road
Warren, MI 48397-5000

Improved Target Acquisition System (ITAS)
Project Manager
CCWS Project Office
ATTN: SFAE-MSL-CWS-F
Building 4505
Redstone Arsenal, AL 35898-5750

Installation Protection Program (IPP) Family of Systems
ATTN: SFAE-CBD-GN
5109 Leesburg Pike
Skyline VI, Suite 916
Falls Church, VA 22041-3247

Integrated Family of Test Equipment (IFTE)
Product Manager–Test, Measurement and Diagnostic Equipment
Building 3651
Redstone Arsenal, AL 35898-5000

Integrated System Control (ISYSCON) (V)4/Tactical Internet Management System (TIMS)
PdM, Network Operations-Current Force
ATTN: SFAE-C3T-NCF
Fort Monmouth, NJ 07703

Interceptor Body Armor (IBA)
ATTN: SFAE-SDR-EQ
10170 Beach Road
Building 325
Fort Belvoir, VA 22060-5800

Javelin
Project Manager, Close Combat Weapon Systems Project Office
ATTN: SFAE-MSL-CWS-J
Redstone Arsenal, AL 35898

Joint Biological Agent Identification Diagnostic System (JBAIDS)
ATTN: SFAE-CBD-CBMS
64 Thomas Johnson Drive
Frederick, MD 21702-5041

Joint Biological Point Detection System (JBPDS)
ATTN: SFAE-CBD-BD
5183 Blackhawk Road
APG, MD 21010-5424

Joint Chemical Agent Detector (JCAD)
ATTN: SFAE-CBD-NBC-D
5183 Blackhawk Road
APG, MD 21010-5424

Joint Combat Identification Marking System (JCIMS)
PM Target Identification & Meteorological Sensors (PM TIMS)
ATTN: SFAE-IEWS-NS-TIMS
Avenue of Memories (563)
Fort Monmouth, NJ 07703

Joint Common Missile (JCM)
Joint Common Missile Project Office
ATTN: SFAE-MSL-CM
5250 Martin Road
Redstone Arsenal, AL 35898-8000

Joint High Speed Vessel (JHSV)
Product Manager
Army Watercraft Systems
ATTN: SFAE-CSS-FP-W
Warren, MI 48397-5000

Joint Land Attack Cruise Missile Defense (LACMD) Elevated Netted Sensor System (JLENS)
PEO Missiles and Space
ATTN: SFAE-MSLS-CMDS-JLN
P.O. Box 1500
Huntsville, AL 35807-3801

Joint Land Component Constructive Training Capability (JLCCTC)
PEO STRI, PM Constructive Simulation
12350 Research Parkway
Orlando, FL 32826-3276

Joint Network Management Systems (JNMS)
PdM, NETOPS-CF
ATTN: SFAE-C3T-NO-CF
Ft. Monmouth, NJ 07703

Joint Precision Airdrop System (JPADS)
Product Manager Force Sustainment Systems, LTC Craig Rettie
508-233-5312
craig.rettie@us.army.mil

Joint Service General Purpose Mask (JSGPM)
Joint Project Manager for Individ-ual Protection
ATTN: JPM-IP
50 Tech Parkway, Suite 301
Stafford, Virginia 22556

Joint Tactical Ground Station (JTAGS)
PEO Air, Space and Missile Defense
ATTN: SFAE-ASMD-AMD-JTG
P.O. Box 1500
Huntsville, AL 35807-3801

Joint Warning and Reporting Network (JWARN)
Space and Naval Warfare Systems Command
4301 Pacific Highway
San Diego, CA 92110

Kiowa Warrior
Product Manager
ATTN: SFAE-AV-AS-ASH
Building 5308
Sparkman Center
Redstone Arsenal, AL 35898

Land Warrior (LW)
PM Sensors and Lasers
ATTN: SFAAE-SDR-SEN
10170 Beach Road
Building 325
Fort Belvoir, VA 22060

Light Utility Helicopter (LUH)
DA Systems Coordinator-Light Utility Helicopter (LUH)
ASA(ALT) Aviation-Intelligence & Electronic Warfare
Attn: SAAL-SAI, Room 10006
2511 South Jefferson Davis Highway
Arlington, VA 22202-3911

Lightweight 155mm Howitzer (LW 155)
ATTN: SFAE-GCS-JLW
Picatinny Arsenal, NJ 7806-5000

Lightweight Laser Designator Rangefinder (LLDR)
Lightweight Laser Designator Rangefinder (LLDR)
ATTN: SFAE-SDR-EQ
10170 Beach Road
Building 325
Fort Belvoir, VA 22060-5800

Lightweight Water Purification (LWP) System
PM Petroleum and Water Systems
6501 East 11 Mile Road
Mail Stop 111
Warren, MI 48397-5000

Line Haul Tractor
ATTN: SFAE-CSS-TV-H
MS 429
6501 East Eleven Mile Road
Warren, MI 48397-5000

Load Handling System Compatible Water Tank Rack (Hippo)
PM Petroleum and Waters Systems
6501 East 11 Mile Road
Mail Stop 111
Warren, MI 48397-5000

Longbow Apache
PM Apache
Building 5681
Redstone Arsenal, AL 35898-5000

M2 Machine Gun Quick Change Barrel Kit
PM Soldier Weapons
(SFAE-SDR-SW)
PEO-Soldier
Picatinny Arsenal, NJ 07806-5000

Maneuver Control System (MCS)
PM GCC2
ATTN: SFAE-C3T-GC-B
Fort Monmouth, NJ 07703

Medical Communications for Combat Casualty Care (MC4)
PM Medical Communications for Combat Casualty Care (MC4)
524 Palacky Street
Fort Detrick, MD 21702-9241

Medium Caliber Ammunition
PM Maneuver Ammunition Systems
ATTN: SFAE-AMO-MAS
Picatinny Arsenal, NJ 07806

Medium Extended Air Defense System (MEADS)
PATRIOT/MEADS Combined Aggregate Program (CAP)
PEO Missiles and Space ATTN: SFAE-MSLS-LT-CAP
P.O. Box 1500
Huntsville, AL 35807-3801

Meteorological Measuring Set–Profiler (MMS-P)
PM Target Identification & Meteorological Sensors (PM TIMS)
ATTN: SFAE-IEWS-NS-TIMS
Avenue of Memories (563)
Fort Monmouth, NJ 07703

Modular Fuel System (MFS)
PM Petroleum and Waters Systems
501 East 11 Mile Road
Mail Stop 111
Warren, MI 48397-5000

Mortar Systems
PM Combat Ammunition Systems
ATTN: SFAE-AMO-CAS-MS
Picatinny Arsenal, NJ 07806

Movement Tracking System (MTS)
Movement Tracking System
PM Logistics Information Systems
800 Lee Avenue, Bldg 5100
Fort Lee, VA 23801-1718

Night Vision Devices
PM Sensors and Lasers
ATTN: SFAAE-SDR-SEN
10170 Beach Road
Building 325
Fort Belvoir, VA 22060

Non-Line of Sight-Launch System (NLOS-LS)
NLOS-LS Project Office
ATTN: SFAE-MSLS-NL
Building 5250, Room B300
Redstone Arsenal, AL 35898-5750

Nuclear Biological Chemical Reconnaissance Vehicle (NBCRV)-Stryker
ATTN: SFAE-CBD-NBC-R
5183 Blackhawk Road
APG, MD 21010-5424

One Semi-Automated Force (OneSAF) Objective System
Product Manager, One Semi Automated Forces
12350 Research Parkway
Orlando, FL 32826-3276

One Tactical Engagement Simulation System (OneTESS)
Project Manager,
Training Devices
12350 Research Parkway
Orlando, FL 32826-3276

Paladin/Field Artillery Ammunition Supply Vehicle (FAASV)
Project Manager HBCT
ATTN: SFAE-GCS-HBCT
6501 East 11 Mile Road
Warren, MI 48397-5000

PATRIOT (PAC-3)
Program Executive Office, Missiles and Space
Lower Tier Project Office
ATTN: SFAE-MSLS-LT
P.O. Box 1500
Huntsville, AL 35807-3801

Precision Guided Mortar Munitions (PGMM)
PM Combat Ammunition Systems
ATTN: SFAE-AMO-CAS-MS
Picatinny Arsenal, NJ 07806

Prophet
PM SW
ATTN: SFAE-IEW&S-SG
Building 288
Sherrill Avenue
Fort Monmouth, NJ 07703

Rapid Equipping Force (REF)
Rapid Equipping Force
10236 Burbeck Road
Fort Belvoir, Virginia 22060

Secure Mobile Anti-Jam Reliable Tactical–Terminal (SMART-T)
PM WIN-T
ATTN: SFAE-C3T-WIN-ESS
Fort Monmouth, NJ 07703-5508

Sentinel
PEO Space and Missile Defense
ATTN: SFAE-MSLS
Redstone Arsenal, AL 35898-5000

Small Arms
PM Soldier Weapons
(SFAE-SDR-SW)
PEO-Soldier
Picatinny Arsenal, NJ 07806-5000

Small Caliber Ammunition
Project Manager Maneuver Ammunition Systems
ATTN: SFAE-AMO-MAS
Picatinny Arsenal, NJ 07806

Sniper Systems
PM Soldier Weapons
(SFAE-SDR-SW)
PEO-Soldier
Picatinny Arsenal, NJ 07806-5000

Stryker
ATTN: SFAE-GCS-BCT MS 325
6501 East Eleven Mile Road
Warren, MI 48397-5000

Surface Launched Advanced Medium Range Air-to-Air Missile (SLAMRAAM)
PEO Air, Space and Missile Defense
ATTN: SFAE-ASMD-SHO
Redstone Arsenal, AL 35898-5000

Tactical Electric Power (TEP)
Tactical Electric Power
10205 Burbeck Road
Fort Belvoir, VA 22060-5863

Tactical Fire Fighting Truck (TFFT)
ATTN: SFAE-CSS-TV-H
MS 429
6501 East Eleven Mile Road
Warren, MI 48397-5000

Tactical Operations Center (TOC)
Tactical Operations Center (TOC)
145 Research Boulevard
Building 12
Madison, AL 35758

Tactical Unmanned Aerial Vehicle (TUAV)
Product Manager
Unmanned Aircraft Systems
ATTN: SFAE-AV-UAS

Tank Ammunition
PM Maneuver Ammunition Systems
ATTN: SFAE-AMO-MAS
Picatinny Arsenal, NJ 07806

Thermal Weapon Sight II
PM Sensors and Lasers
ATTN: SFAAE-SDR-SEN
10170 Beach Road
Building 325
Fort Belvoir, VA 22060

Transportation Coordinators' Automated Information for Movement System II (TC-AIMS II)
PM TIS
8000 Corporate Court
Springfield, VA 22153

Tube-Launched, Optically-Tracked, Wire-Guided (TOW) Missiles
Project Manager, CCWS Project Office
ATTN: SFAE-MSL-CWS-F
Building 4505
Redstone Arsenal, AL 35898-5750

Unified Command Suite (UCS)
ATTN: SFAE-CBD-WMDCSS
5183 Blackhawk Road
APG, MD 21010-5424

Unit Water Pod System (Camel)
PM Petroleum and Water Systems
ATTN: LTC Michael Receniello
6501 East 11 Mile Road
Mail Stop 111
Warren, Michigan 48397-5000

Warfighter Information Network–Tactical (WIN-T)
Project Manager, WIN-Tactical
ATTN: SFAE-C3T-WIN
Building 909
Murphy Dr.
Fort Monmouth, NJ 07703

XM101 Common Remotely Operated Weapon Station (CROWS)
PM Soldier Weapons
(SFAE-SDR-SW)
PEO-Soldier
Picatinny Arsenal, NJ 07806-5000

XM307
PM Soldier Weapons
(SFAE-SDR-SW)
PEO-Soldier
Picatinny Arsenal, NJ 07806-5000

Design & Layout: Integrated Communications Team, SAIC